AF389385

Industrial Energy Management and Sustainability

Industrial Energy Management and Sustainability

Editors

Miriam Benedetti
Vito Introna

MDPI • Basel • Beijing • Wuhan • Barcelona • Belgrade • Manchester • Tokyo • Cluj • Tianjin

Editors
Miriam Benedetti
Energy Efficiency Department,
Italian National Agency for New Technologies,
Energy and Sustainable Economic Development,
Italy

Vito Introna
University of Rome Tor Vergata
Italy

Editorial Office
MDPI
St. Alban-Anlage 66
4052 Basel, Switzerland

This is a reprint of articles from the Special Issue published online in the open access journal *Sustainability* (ISSN 2071-1050) (available at: www.mdpi.com/journal/sustainability/special_issues/ Industrial_Energy_Management).

For citation purposes, cite each article independently as indicated on the article page online and as indicated below:

LastName, A.A.; LastName, B.B.; LastName, C.C. Article Title. *Journal Name* **Year**, *Volume Number*, Page Range.

ISBN 978-3-0365-1746-9 (Hbk)
ISBN 978-3-0365-1745-2 (PDF)

Contents

About the Editors

Miriam Benedetti is a researcher at ENEA, the Italian National Agency for New Technologies, Energy and Sustainable Economic Development. Her current work focuses on the identification of solutions to overcome technological, economic, organisational and information barriers to the uptake of the best available technologies and best practices for energy efficiency in industry. She is the project manager of the project "Energy efficiency of industrial products and processes"(EUR 8.6 M funded by the Italian Ministry for Economic Development) and her research activities particularly focus on the creation of support tools to help industrial companies identify and implement waste heat recovery opportunities. She was previously a senior researcher at EURAC (Italy) and research associate at the University of Cambridge (UK). She holds a PhD in Industrial Engineering and an MSc in energy engineering from the University of Rome Tor Vergata.

Vito Introna is an associate professor at the University of Rome Tor Vergata where he teaches "Innovation and Project Management". He received his PhD in Industrial Product and Process Engineering at the University of Naples "Federico II"in 2003. Since 2000, he has been conducting a research project for national and international research centers and international companies and he has held courses within Master programs and certification programs.

Preface to "Industrial Energy Management and Sustainability"

Growing environmental concerns caused by increasing consumption of natural resources and pollution need to be addressed. Manufacturing dictates the efficiency with which resource inputs are transformed into economically valuable outputs in the form of products and services. Consequently, it is also responsible for the resulting waste and pollution generated from this transformation process. As a matter of fact, about one-third of the global total energy consumption is associated with manufacturing activities; thus, achieving a higher energy efficiency in this sector has been the focus of research as well as of policy and industrial programmes in recent years. In particular, being able to effectively manage energy and energy-related activities has proved to be a fundamental capability for companies willing to improve their sustainability, as it constitutes the first, critical step to understanding their processes and to identifying and correctly evaluating improvement opportunities.

This Special Issue focuses on energy management and sustainability of both manufacturing processes and systems, including methods, practices, tools, applications and experiences. It is particularly intended for academics willing to gather information on relevant work going on this area and valuable data from different industrial sectors as well as for practitioners who wish to be updated on latest practices and methodologies, thus facilitating knowledge transfer. The editors would like to warmly thank all contributing authors.

Miriam Benedetti, Vito Introna
Editors

Editorial

Industrial Energy Management and Sustainability

Miriam Benedetti [1,*] and Vito Introna [2,*]

[1] Italian National Agency for New Technologies, Energy and Sustainable
     Economic Development (ENEA), 00123 Rome, Italy
[2] Department of Enterprise Engineering, University of Rome Tor Vergata, 00133 Rome, Italy
*   Correspondence: miriam.benedetti@enea.it (M.B.); vito.introna@uniroma2.it (V.I.)

Citation: Benedetti, M.; Introna, V. Industrial Energy Management and Sustainability. *Sustainability* **2021**, *13*, 8814. https://doi.org/10.3390/su13168814

Received: 3 August 2021
Accepted: 5 August 2021
Published: 6 August 2021

Publisher's Note: MDPI stays neutral with regard to jurisdictional claims in published maps and institutional affiliations.

Increasing the sustainability of industrial activities is a top priority for national and supranational governmental institutions. Given both their resource intensity and their criticality for competitiveness, industries of all sectors have been at center stage in the fight against pollution and global warming over the last decades as well as the focus of research and of policy and industrial programs in recent years. In this context, energy management and energy efficiency opportunities and technologies play an important role, as energy is a particularly critical resource in terms of cost, availability, and intensity. Despite the increasing attention paid to this topic by academics and policy makers, it is often noted that practitioners and undertakings still present implementation issues and demand support tools and updated data in order to facilitate the identification of practices and technologies, benchmark activities, knowledge transfer, and, ultimately, the transition to more sustainable business models and production systems. Such demands have also become more urgent in the last year, when the COVID-19 pandemic demonstrated that resilience and flexibility are necessary features for a business to thrive and are intimately related to the ability to efficiently and effectively manage resources and energy, in particular.

This Special Issue focuses on Energy Management and Sustainability of both manufacturing processes and systems, including methods, practices, tools, applications, and experiences. It aims on the one hand at highlighting recent advances in the field and, on the other hand, at proposing ready-to-use tools and data that can be valuable for practitioners willing to better understand the topic and contextualize it to their sector and business. It includes a focus on a specific and critical energy efficiency opportunity such as waste heat recovery with the proposal of a database where companies can gather valuable ideas and data (Contribution 1); three in-depth analyses of specific sectors and technologies (CHP in the ceramic sector for Contribution 2, cement industry for Contribution 3, and metal extraction for Contribution 4), all supported by careful and robust data collection and analyses; and the proposal of a broadly applicable methodology to translate energy management theory into practice, providing effective and practical support. Some of the papers (Contribution 2, Contribution 3, and Contribution 4, in particular) are based on data referring to a specific geographical area, but we think that information and outcomes can be considered of general validity and are certainly a fundamental starting point for replication and application to different contexts.

Benedetti et al. (Contribution 1) focus on industrial waste heat recovery, which is nowadays considered one of the hot topics when it comes to energy efficiency and resource preservation. In their paper, a methodology is presented for waste heat recovery opportunities identification as well as two distinct databases containing waste heat recovery case studies and technologies. The databases can be considered as a tool to enhance knowledge transfer in the industrial sector. Through an in-depth analysis of the scientific literature, the two database structures were developed to define the fields and information to collect, and then preliminary population was performed. To highlight the usability of research outcomes by practitioners, a validation phase was carried out and main results are presented.

1

Branchini et al. (Contribution 2) present the preliminary results of a research project aimed at defining the benefits of using combined heat and power (CHP) systems in the ceramic sector. Their study is based on data collected from ten CHP installations in Italian ceramic plants, which allowed them to outline the average characteristics of prime movers and to quantify the contribution of CHP thermal energy in supporting the dryer process. Data revealed that when the goal is to maximize the generation of electricity for self-consumption, internal combustion engines are the preferred choice because of their higher conversion efficiency. In contrast, gas turbines allowed minimizing the consumption of natural gas by the spray dryer.

Cantini et al. (Contribution 3) describe the recent application of energy efficiency solutions and technologies in the Italian cement industry and their future perspectives. They analyzed a sample of plants by considering the type of interventions they recently implemented or intend to implement. The outcome is a descriptive analysis, useful for companies willing to improve their sustainability. Results prove that measures to reduce the energy consumption of auxiliary systems such as compressors, engines, and pumps are currently the most attractive opportunities. Moreover, the results prove that consulting with sector experts enables the collection of updated ideas for improving technologies, thus giving valuable inputs to scientific research.

Imasiku and Thomas (Contribution 4) present an evaluation of energy efficiency opportunities in copper operations and the environmental impact of metal extraction by means of a case study on the Central African Copperbelt countries of Zambia and the Democratic Republic of Congo. In addition, four strategies are identified by which the mining and technology industries can enhance sustainable electricity generation capacity: energy efficiency; use of solar and other renewable resources; sharing expertise from the mining and technology industries within the region; and taking advantage of the abundant cobalt and other raw materials to initiate value-added manufacturing.

Solnørdal and Nilsen (Contribution 5) explore the implementation of a corporate environmental program in an incumbent firm and the ensuing emergence of energy management practices. Translation theory and the "travel of management ideas" are used as a theoretical lens in this case study when analyzing the process over a 10-year period. Furthermore, based on a review and synthesis of prior studies, a "best Energy Management practice" is developed and used as a baseline when assessing the energy management practices of the case firm.

The editors would like to thank all authors for their valuable contributions and MDPI for all the support in putting it together. We hope that readers can find information, data, and inspiration to keep up the good work on industrial energy efficiency and sustainability.

Author Contributions: All authors contributed to the idea and the design of the editorial; M.B. prepared the original draft; V.I. contributed to the review and editing. All authors have read and agreed to the published version of the manuscript.

Funding: The research activity received no external funding.

Conflicts of Interest: The authors declare no conflict of interest.

List of Contributions:

1. Benedetti, M.; Dadi, D.; Giordano, L.; Introna, V.; Lapenna, P.E.; Santolamazza, A. Design of a Database of Case Studies and Technologies to Increase the Diffusion of Low-Temperature Waste Heat Recovery in the Industrial Sector.
2. Branchini, L.; Bignozzi, M.C.; Ferrari, B.; Mazzanti, B.; Ottaviano, S.; Salvio, M.; Toro, C.; Martini, F.; Canetti, A. Cogeneration Supporting the Energy Transition in the Italian Ceramic Tile Industry.
3. Cantini, A.; Leoni, L.; De Carlo, F.; Salvio, M.; Martini, C.; Martini, F. Technological Energy Efficiency Improvements in Cement Industries.
4. Imasiku, K.; Thomas, V.M. The Mining and Technology Industries as Catalysts for Sustainable Energy Development.
5. Solnørdal, M.T.; Nilsen, E.A. From Program to Practice: Translating Energy Management in a Manufacturing Firm.

 *sustainability*

Article

Design of a Database of Case Studies and Technologies to Increase the Diffusion of Low-Temperature Waste Heat Recovery in the Industrial Sector

Miriam Benedetti [1,*], Daniele Dadi [2,*], Lorena Giordano [1], Vito Introna [2], Pasquale Eduardo Lapenna [1,3] and Annalisa Santolamazza [4]

[1] Italian National Agency for New Technologies, Energy and Sustainable Economic Development (ENEA), 00123 Rome, Italy; lorena.giordano@enea.it (L.G.); pasquale.lapenna@uniroma1.it (P.E.L.)

[2] Department of Enterprise Engineering, University of Rome Tor Vergata, 00133 Rome, Italy; vito.introna@uniroma2.it

[3] Department of Mechanical and Aerospace Engineering, University of Rome "La Sapienza", 00184 Rome, Italy

[4] DEIM School of Engineering, University of Tuscia, 01100 Viterbo, Italy; annalisa.santolamazza@unitus.it

* Correspondence: miriam.benedetti@enea.it (M.B.); daniele.dadi@uniroma2.it (D.D.)

Abstract: The recovery of waste heat is a fundamental means of achieving the ambitious medium- and long-term targets set by European and international directives. Despite the large availability of waste heat, especially at low temperatures (<250 °C), the implementation rate of heat recovery interventions is still low, mainly due to non-technical barriers. To overcome this limitation, this work aims to develop two distinct databases containing waste heat recovery case studies and technologies as a novel tool to enhance knowledge transfer in the industrial sector. Through an in-depth analysis of the scientific literature, the two databases' structures were developed, defining fields and information to collect, and then a preliminary population was performed. Both databases were validated by interacting with companies which operate in the heat recovery technology market and which are possible users of the tools. Those proposed are the first example in the literature of databases completely focused on low-temperature waste heat recovery in the industrial sector and able to provide detailed information on heat exchange and the technologies used. The tools proposed are two key elements in supporting companies in all the phases of a heat recovery intervention: from identifying waste heat to choosing the best technology to be adopted.

Keywords: waste heat recovery; low-temperature waste heat; database development; industrial sustainability

Citation: Benedetti, M.; Dadi, D.; Giordano, L.; Introna, V.; Lapenna, P.E.; Santolamazza, A. Design of a Database of Case Studies and Technologies to Increase the Diffusion of Low-Temperature Waste Heat Recovery in the Industrial Sector. *Sustainability* **2021**, *13*, 5223. https://doi.org/10.3390/su13095223

Academic Editor: Enrique Rosales-Asensio

Received: 30 March 2021
Accepted: 3 May 2021
Published: 7 May 2021

Publisher's Note: MDPI stays neutral with regard to jurisdictional claims in published maps and institutional affiliations.

1. Introduction and Background

The European Union (EU) is fighting climate change through ambitious policies. Reductions in greenhouse gas emissions have reached the 2020 target. The new European Commission proposal to cut greenhouse gas emissions by at least 55% by 2030 sets Europe on a responsible path to becoming climate neutral by 2050.

One of the most significant contributions towards meeting the EU 2050 target is to expand the use of renewable energy solutions: an increased optimization and development in hydropower [1,2], wind power [3–5], solar energy [6,7], energy storage [8–11], and an increased share of these technologies in the energy mix are essential for a sustainable energy transition.

An equally important contribution is the improvement in energy efficiency, especially in those sectors with high energy consumption.

Waste heat recovery (WHR) is a key element of this endeavour. It consists of recovering heat from waste energy flows and reusing it directly or converting it into other forms of energy. Many sectors are involved, in particular the industrial sector, the transport sector, the power generation sector, and the urban sector (urban WHR), which are all characterized by a large availability of waste heat and still unexploited potential [12].

EU policies have highlighted the evidence of this opportunity, thanks to which several European projects such as Einstein and Einstein II [13,14], Greenfoods [15], and sEEnergies [16] have been developed. In addition to the development of useful support tools, all these projects have had the great merit of emphasizing the unexpressed potential of waste heat recovery.

From the need to reduce the gap between the availability of waste heat and its recovery rate, several important recent studies have focused on different aspects of heat recovery: quantification and estimation of available waste heat [12,17], perspectives of matching supply demand [18], analysis and development of technologies for heat recovery [19], and the opportunities for reusing the heat recovered. Concerning this final aspect, two main ways have been identified to reuse the recovered heat: internal uses, when the recovered heat is reused inside the system in which the heat is recovered, or external uses, where the recovered heat is used externally, such as industrial symbiosis [20] and district heating networks [21,22], for example from urban waste heat. An important contribution to achieving the 2050 climate goal can derive from both internal and external improvements, considering the availability of several options to utilize these unused heat resources [21].

One of the most important factors in evaluating the feasibility of a waste heat recovery intervention depends on the temperature at which the heat is available. The temperature determines the quality of waste heat and strongly influences both the applicable technologies and the economic practicability of its recovery.

There are no particular barriers to heat recovery at high temperatures: this process is more feasible, mainly due to the availability of more mature technologies and the greater energetic efficiencies involved, which results in a more immediate economic return.

As regards WHR at low temperatures, commonly defined as the waste heat available at a temperature lower than 200–250 °C [23], the situation is different; due to its low exergy, low-grade waste heat is more difficult to use [24].

The industrial sector boasts one of the greatest availabilities of waste heat at low temperatures. According to [24], 66% of the total waste heat in the industry is available at a temperature below 200 °C.

To make the most of this important resource, in the last decade, many low-grade heat recovery technologies have been developed. Nowadays, the market offers various consolidated technologies such as organic Rankine Cycles (ORC), heat pumps (HP), various heat exchangers, and many other technologies under development.

Nevertheless, the number of WHR applications in the industrial sector is still limited, even though industrial waste heat at low temperatures is abundantly available and the technologies for its use are mature. This misalignment between supply and demand is primarily due to technical, informative, economic, and organizational barriers [18,25,26]. In this regard, Figure 1 shows a schematic representation of the most frequent barriers found in the literature to low-grade waste heat recovery.

Among the various barriers identified, the lack of knowledge and the absence of support tools significantly influence WHR technologies' diffusion [27–29].

The work presented in this article aims to create support tools focused on heat recovery at low temperatures in the industrial sector. It is part of a three-year project aiming to increase the spread of WHR applications in the industrial scenario by preventing non-technological barriers. This article focuses on the design and construction of two databases regarding technologies and case studies for low-temperature WHR. These tools represent two key elements of an information system developed within the scope of a broader research project, whose final aim is to develop a software tool supporting the industrial companies in all phases involved in the planning of WHR projects, from the identification of waste heat flows to the selection of the best technological solution for its reuse.

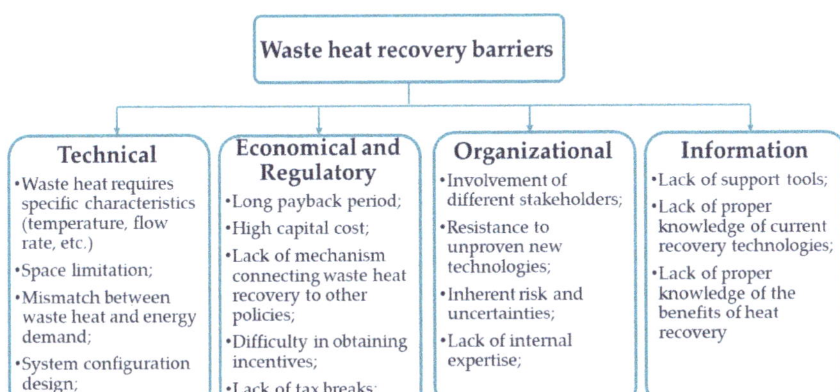

Figure 1. Technical, economic, regulatory and information barriers to low-grade waste heat recovery and utilization [18,25,26].

In recent years, other projects have developed databases aimed at overcoming non-technical barriers to the diffusion of energy efficiency measures. EU-MERCI [30] is a Horizon 2020 European project aimed at spreading the existing knowledge about energy efficiency in the industry. One of the most significant outputs of the EU-MERCI project is developing a platform that makes available a database containing energy efficiency projects. The database contains almost three thousand projects put in place in different EU countries [31].

Another tool which has been proposed by the Industrial Assessment Centers (IAC) is an available online database containing a collection of all the publicly available industrial energy assessment and recommendation data. The database includes information on the type of facility assessed (size, industry, energy usage, etc.) and details of resulting recommendations (type, energy and economic savings, etc.) [32]. In March 2021, the IAC database contains 19,444 assessments (from United States manufacturer) and 146,405 recommendations.

The MAESTRI project [33] intends to develop an innovative and integrated platform combining holistic efficiency assessment tools, a novel management system, and an innovative approach for industrial symbioses implementation. One of its most significant outputs is the creation of a waste database to support the identification of the potential (re)use of waste energy flows.

The Solar Heat for Industrial Processes (SHIP) database [34] has been created in the IEA Task 49/IV framework. This online database contains a worldwide overview of existing solar thermal plants, which provide thermal energy for production processes for different industrial sectors. Each record contains information about the collector field's size, collector technology, or integration point in the production process.

The "Matrix of Industrial Process Indicators" is an online database constantly being developed in various projects by the AEE—Institute for Sustainable Technologies (AEE IN-TEC) advisors [35]. The database is designed as a decision support tool to help the industry concerning energy efficiency and to aid the identification of suitable solar applications.

Although the illustrated tools are dense with information, they will not completely overcome all the gaps identified above for WHR technologies' diffusion. IAC, EU-MERCI, and "Matrix of Industrial Process Indicators" databases contain numerous applications but do not provide detailed information on the technologies used. Furthermore, IAC and EU-MERCI databases are based only on a subset of projects and refer to a single specific geographical area: the United States and Europe, respectively. The MAESTRI and the SHIP databases are focused only on specific applications such as external uses of waste heat and solar thermal plant recovery applications. All these databases are not exclusively focused on WHR.

Our objective was to create innovative, complete, and easy-to-use tools able to overcome the gaps still not completely resolved by those previously developed. The integrated tools proposed in this article are the first example in the literature of databases completely focused on low-temperature WHR in the industrial sector. They are built to provide companies with the basis for a comprehensive assessment of their possibilities in heat recovery. They include detailed information on heat exchange and the technologies used. They have been designed to be populated with data from different sources, including inspections, interviews, energy audits, catalogs, and scientific articles. Compared to the other tools present in the literature, they are not specific to a single geographical area.

The paper discusses the design and the implementation of two different databases supporting the decision-making process for implementing WHR systems in industrial facilities: the *Technology database* and the *Case study database*. The former provides an insight into both well-established and emerging technologies for low-temperature WHR, while the latter analyzes some of the best practice examples of WHR in industrial processes. Although the two databases are conceived as separate tools, they can also interact with each other due to a common structure.

The following sections describe all the phases that led to the development of the two different tools. Moreover, an analysis of the preliminary population results and a practical example of their possible use are reported.

2. Database Design Methodology

The first step in developing the databases was the definition of the state-of-the-art of low-temperature waste heat recovery technologies and applications in industrial processes.

On a preliminary basis, we focused our attention on the classification of waste heat depending on temperature levels, in order to set the quantitative boundaries of the study and to define the scope of the databases.

Different classifications of waste heat according to temperature ranges can be found in technical literature [36,37]. In this work, reference was made to a classification based on five temperature ranges proposed by the U.S. Department of Energy (DOE) by [23] and summarized in Table 1.

Table 1. Classification of waste heat according to temperature level [23].

Range of Waste Heat	Temperature
Ultra-low temperature	<120 °C
Low temperature	120–230 °C
Medium temperature	230–650 °C
High temperature	650–870 °C
Ultra-high temperature	>870 °C

As previously stated, in line with the research project's objectives, attention was focused on ultra-low and low-temperature waste heat.

The next step concerned the design and development of the databases: the *Technology database* and the *Case study database*. As already highlighted, they have been designed as two distinct tools, but in order to facilitate the data exchange and, thus, the integrated use of the databases, a standard procedure was set up for their implementation.

The procedure is divided into four macro-phases:

1. *Analysis of the reference context*: preliminary analysis of the reference scenario and formulation of the research keywords.
2. *Analysis of the scientific literature*: search and preliminary analysis of the literature sources, to select the most representative studies to be used as input for the definition of the database structure.

3. *Definition of the database structure*: in this phase, thanks to the results of the literature analysis, the database fields and the information to be collected are defined.
4. *Population of the database*: the last phase involves further analysis of the sources found and their use for the database's preliminary population.

Figure 2 schematically shows the main phases involved in the database development process.

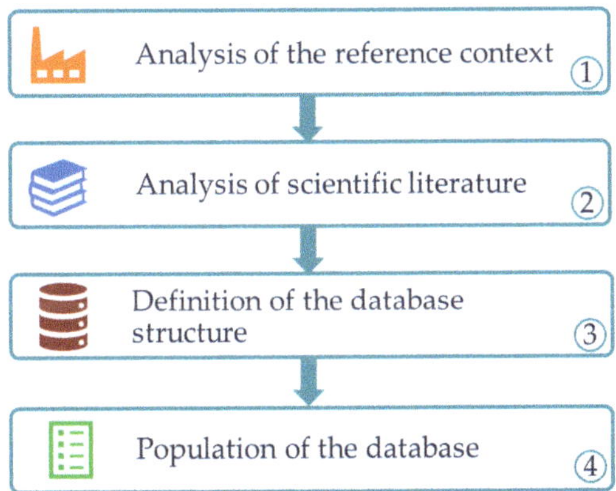

Figure 2. Graphical representation of the main phases for the creation of the two databases.

For both databases, the analysis of the context and the bibliographic research carried out provided the fundamental inputs for implementing the subsequent phases.

In order to define the preliminary structure, the articles revealed during the bibliographic research conducted were analyzed. Among the available sources, those containing the most significant amount of information on WHR technologies at low and ultra-low temperatures and their application in heat recovery from industrial processes were selected. The detailed analysis of these references made it possible to understand what information must be collected to adequately characterize WHR technologies and applications. In addition, similar existing databases from the literature were taken as a reference, in particular, the MAESTRI database [33] concerning industrial symbiosis (a WHR process is similar in terms of information to an exchange of materials, energy, or products between different industrial plants), and the EUMERCI database [30] containing energy efficiency case studies (including WHR interventions). Although focused on different topics, these tools have common traits in terms of type of information contained.

After defining the structure of the databases, all the sources detected were further analyzed to be used for the preliminary population of the two tools. A first evaluation was carried out, matching the document's content with the project's boundaries:

*   WHR case studies and technologies at ultra-low and low temperatures (temperature lower than 230 °C);
*   Relevance and applicability to the industrial sector (for example, applications in the residential sector were not considered).

Then, a further screening was carried out to identify the studies in which most of the information required by the database fields was available (at least information relating to the waste heat characterization, applicability of the technologies, and reuse of the heat recovered); finally, data were collected and used for database population.

After performing the four steps, both databases were subjected to a validation procedure, thanks to the interaction with important stakeholders, such as WHR technology producers and companies operating in the industrial sector as possible users of the tools. The validation procedure mainly concerned three aspects:

1. Test the effectiveness of the two databases in containing all the information necessary to adequately characterize the various technologies (*Technology database*) and applications (*Case study database*);
2. Verify the correctness of the data contained in the databases and evaluate the correspondence of the information collected, in terms of technologies available and effective applications in the industrial scenario;
3. Test the usability of the tools in their role of supporting the identification of heat recovery opportunities.

The databases' development, which is presented in the following section, results from several interactions. In fact, the numerous feedbacks received through the interaction with the main stakeholders were then used to make the appropriate changes to the structure of the two databases.

3. The Technology Database

3.1. Analysis of the Reference Context

As suggested by [36], technologies used to recover waste heat from industry can be categorized depending on how the waste heat is reused: it can be used directly (at the same or at a lower temperature level), or it can be transformed to another form of energy or to a higher temperature.

We can identify four macro-categories:

1. *Waste Heat to Heat*: technologies through which the waste heat recovered is used to produce thermal energy at a higher temperature level (heat pumps, mechanical vapor compression, absorption heat pumps, etc.).
2. *Waste Heat to Cold*: technologies that utilize the waste heat recovered to produce cooling energy (absorption and adsorption chiller, etc.).
3. *Waste Heat to Power*: technologies that convert the waste heat recovered to electricity (organic Rankine cycles, Kalina cycles, etc.).
4. *Heat Exchange*: technologies through which the waste heat recovered is used directly at the same or a lower temperature. Heat exchangers and thermal energy storage are the two dominant technologies of this category.

To define an appropriate structure of the *Technology database*, it was first necessary to delineate a detailed state-of-the-art regarding the technologies for low-grade WHR currently available. Thanks to the analysis of some important scientific literature contributions [18,19,23,26,36], it was possible to define a general framework of both consolidated (typically adopted in the industry) and emerging (developed and tested at the laboratory or pilot scale) technologies.

Table 2 shows some of the main technologies divided by type of recovery and temperature level within the research project's boundaries.

Table 2. Consolidated and emerging WHR technologies by temperature range [18,19,23,26,36].

Category	Consolidated Technologies		Emerging Technologies	
	<120 °C	120–250 °C	<120 °C	120–250 °C
Waste heat to heat	Heat pumps Absorption HP [1]	Heat pumps	Thermoacoustic HP [1] High temperature HP [1]	Heat transformer High temperature HP [1]
Waste heat to cold	Absorption chiller	-	Adsorption chiller	Thermoacoustic Chiller
Waste heat to power	ORC [2]	ORC [2] Kalina cycle	Electrochemical systems Piezoelectric and pyroelectric systems	Thermoelectric generator

Table 2. *Cont.*

Category	Consolidated Technologies		Emerging Technologies	
	<120 °C	120–250 °C	<120 °C	120–250 °C
Heat Exchange	Shell, tube and plate heat exchangers	Shell, tube and plate heat exchangers	Non-metallic and corrosion-resistant heat exchangers	Recuperators with innovative geometries
	Air preheaters	Heat pipe exchanger	Membrane type systems for latent heat recovery	Advanced design of metallic heat wheel heat pipe exchanger
	Direct contact water heaters	Metallic heat wheel	Desiccant systems for latent heat recovery	Self-recuperative burners
	Non-metallic heat exchangers	Convection recuperator	Systems with phase change material	Systems with phase change material

[1] Heat pump (HP); [2] Organic Rankine cycle (ORC).

Among these technologies, organic Rankine cycles, heat pumps, absorption chillers, and the various types of heat exchangers are the most widespread and consolidated WHR technologies in the industrial scenario. Figure 3 shows the subdivision into categories of the WHR technologies and, for each of them, the most representative technology.

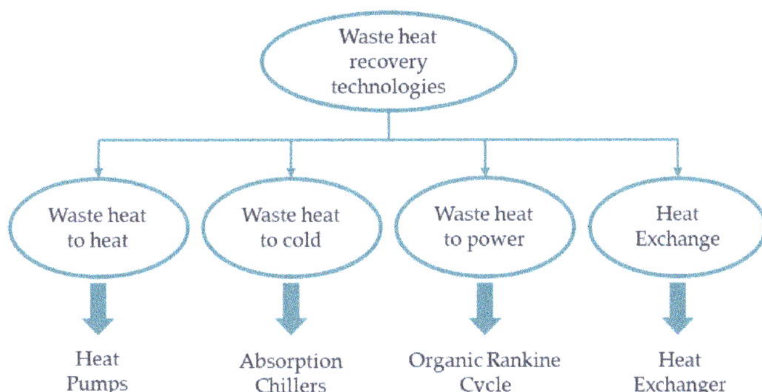

Figure 3. Categorization of waste heat recovery technologies. Elaborated from [36].

3.2. Analysis of the Scientific Literature

A more specific analysis of the literature was conducted to provide input to the definition of the database structure. The literature review, described in detail in [38], was carried out consulting the Scopus online database using the combination of keywords "heat recovery", "low temperature", and "industrial" (in article title, abstract, and keywords). The search produced more than 500 results. A first evaluation was done matching the document's content with the project's boundaries (ultra-low and low temperatures applications) and analyzing the relevance to the industrial sector. This activity identified about 300 articles that were deeply analyzed.

3.3. Definition of the Database Structure

The literature analysis allowed for the identification of the most appropriate database structure to develop a comprehensive, but easy-to-use, tool.

The database is organized as follows: each row of the database corresponds to a low-temperature heat recovery technology. There can be multiple records for each source (for example, a supplier may have several technologies in its catalog).

The information collected is organized into five categories:

1. Identification: includes fields to locate the technology within the database. Different technologies can be associated with each technology provider.

2. Source: contains information relating to the source from which the technology was identified to guarantee its retrieval.
3. Technology provider: contains the information necessary to identify the manufacturer of the technology under investigation.
4. Technology information: identifies the technology, its state of maturity, and the type of recovery it allows to obtain. It also contains the main technical characteristics and ranges of application of this technology.
5. Other information: includes other technical or economic information not recorded in the previous fields that may help the user evaluate and understand the technology.

Table 3 shows all the database fields and their description.

Table 3. *Technology database*'s fields and description.

Category	Field	Description
Identification	Provider ID	A unique progressive number assigned to the WHR[1] provider technology
	Technology ID	A unique progressive number assigned to the WHR[1] specific technology
Source	Source type	Scientific article, project report, catalog, interview, etc.
	Provider/Authors	Name of the technology provider or authors
	Link	Source reference (e.g., DOI[2], website, etc.)
	Year	Time reference (e.g., year of publication, year of the catalog, year of the interview with the supplier, etc.)
Technology provider	Geographical reference	The country or countries in which the technology provider works
	Contact info	Useful information to contact the technology provider (e.g., email address)
	Type of provider	Manufacturer, retailer or both
Technology information	Recovery type	Indicates the type of heat recovery (e.g., waste heat to electricity, waste heat to cooling, etc.)
	Technology	type of technology (e.g., ORC[3], heat pumps, etc.)
	Technology description	Additional information on the technology
	State of maturity	It represents the state of maturity of the technology (e.g., consolidated, emerging, etc.)
	Applications	Description of the possible primary applications of the technology
	Model	If the technology can be identified with a specific model
	Vector fluid	Vector fluid used in the recovery process (e.g., water, oil, etc.)
	Input temperature	Range of admissible input temperatures of the vector fluid
	Minimum input temperature	Minimum admissible input temperatures of the vector fluid
	Vector fluid flow rate	Range of vector fluid flow rate that can be processed

Table 3. *Cont.*

Category	Field	Description
	Working fluid	Working fluid used by technology (e.g., hydrocarbons, refrigerants, etc.)
	Input thermal power	Range or nominal input power
	Output type	Type of output energy vector (e.g., electricity, thermal energy, cooling energy)
	Output power	Range of nominal output power
	Output temperature	Range of output temperatures of the vector fluid
	Efficiency	Characteristic performance parameter of the technology (e.g., efficiency, COP [4], etc.)
	Dimension	Size required by the technology (e.g., m^2/kWh)
Other information	Expected lifetime	The useful life for which the technology was designed
	Expected PBP [5]	Typical payback time
	Notes	Other relevant information about the WHR [1] technology

[1] Waste heat recovery (WHR); [2] Digital object identifier (DOI); [3] Organic Rankine cycle (ORC); [4] Coefficient of performance (COP); [5] Payback period (PBP).

The final structure proposed is the result of several iterations. Initially, only closed fields were foreseen, but it was preferred to maintain a more flexible structure, thanks to the inputs obtained from the interaction with the technology suppliers. The database is designed to contain both single models or model series and product range as, in many cases, the manufacturers offer application-specific designs.

For this reason, it was preferred for some fields to provide a range of value or free completion.

3.4. Population of the Database

The *Technology database*'s preliminary population was carried out using the sources of literature collected during the bibliographic research that was previously carried out. Other information was collected by directly consulting the catalogs of suppliers available online and carrying out interviews with some of the main waste heat technology providers operating in the Italian market.

There are 21 sources used for the *Technology database* population, for a total of 62 technologies. Figure 4a shows that the identified technologies primarily derive from supplier catalogs (77%) and from the interviews conducted (18%). Figure 4b shows the division into categories of the technologies identified.

As shown in Figure 5, organic Rankine cycles (34%), heat pumps (27%), and absorption chillers (15%) are the most numerous technologies identified.

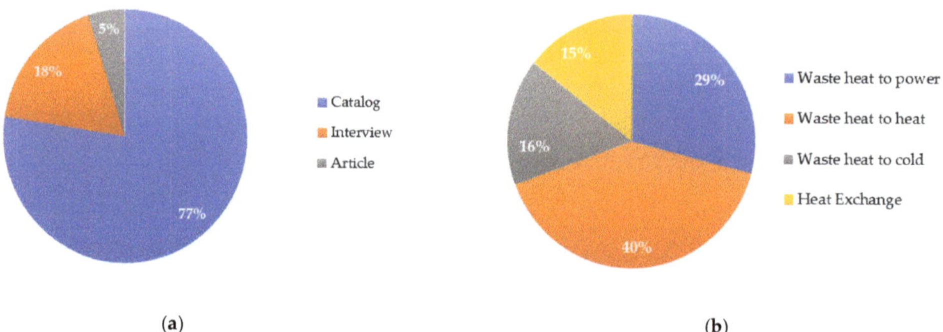

(a)   (b)

Figure 4. Results of the *Technology database*'s preliminary population: (**a**) Type of source analyzed; (**b**) Waste heat recovery type.

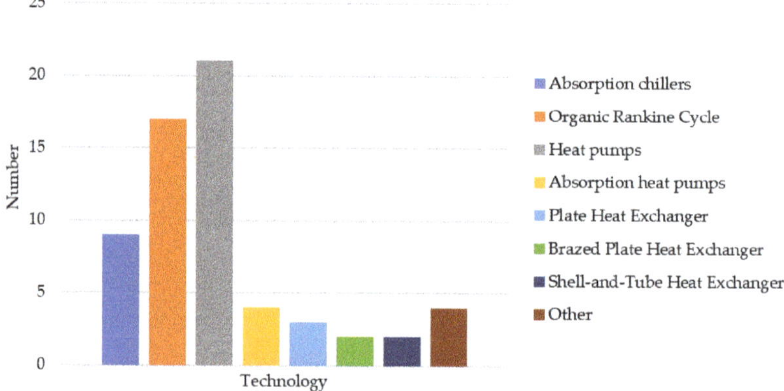

Figure 5. The technologies identified during the *Technology database*'s preliminary population.

4. The Case Study Database

4.1. Analysis of the Reference Context

A preliminary analysis of a limited number of documents [23,26,39] was carried out to identify the industrial sectors with high potential for waste heat recovery at medium–low temperature levels. In this regard, Table 4 provides an overview of the most common waste heat sources and the corresponding temperature levels according to the type of industrial sector [26,39].

Table 4. Industrial processes with low-temperature waste heat production [26,39].

Industrial Sector	Low-Grade Waste Heat Source	Temperature (°C)
	Stack gas from crude distillation	156
Petrochemical	Stack gas from vacuum distillation	216
	Exhaust from ethylene furnace	149
	Waste gas from coke oven	200
Iron/steel making	Blast furnace gas	450
	Exhaust gases from Cowper regenerators	250
	Exhaust gases from electric arc furnaces	204

Table 4. *Cont.*

Industrial Sector	Low-Grade Waste Heat Source	Temperature (°C)
Aluminum	Exhaust from aluminum casting with a stack melter	121
Food and drink	Extracted air from cooking with fryers or ovens	150–200
	Exhaust from drying with spray/rotary dryers	110–160
	Water vapor from evaporation and distillation	100
Textile	Dyed waste water from drying	90–94
	Stenter exhaust for fabric drying and finishing	180
	Waste water rejected from heat exchangers	58–66
Paper	Waste steam from slag flushing in furnace	95–100
	Waste water from slag flushing in furnace	65–85
	Cooling water from furnace wall cooling	35–45
Cement	Exhaust from cement kilns using 5- or 6-stage preheaters	204–300
	Hot air discharged from clinker coolers	100
Chemical and Pharmaceutical	Distillation processes	100–300

Furthermore, the analysis of these sources allowed for the identification of "keywords" (see Table 5) to be used for bibliographic research on main scientific and non-scientific databases and search engines (Scopus, ScienceDirect, Google Scholar, Google). Documents identified by such research included scientific articles, project reports, and studies related to recovery projects already implemented or in the implementation phase.

Table 5. Keywords used in bibliographic research for the *Case study database* development.

Focused on the Research Subject	Focused on Technologies	Focused on Sectors with High Waste Heat Recovery Potential
Waste heat recovery, low-temperature waste heat, low grade industrial waste heat, low grade industrial waste heat case study, industrial waste heat recovery case study, energy recovery, industrial energy symbiosis, industrial symbiosis	Heat pump, Organic Rankine Cycle, Heat Exchanger Network	Agrifood, Dairy industry, Food, Paper and Pulp Industry, Chemical Industry, Textile industry

4.2. Analysis of the Scientific Literature

A preliminary selection was carried out to identify studies focusing on waste heat recovery at temperatures compatible with the project's objectives. At the end of this first selection, the total number of documents to be considered for further analysis was 130.

Among these studies, a few articles (reporting the greatest amount of information on the application of waste heat recovery solutions in industrial processes) were selected to define the preliminary structure of the *Case study database*. In addition, similar existing databases from the literature were taken as a reference, such as those regarding case studies on industrial symbiosis (i.e., MAESTRI [33]). Indeed, in terms of structure and type of information to be collected, a process of waste heat recovery (internal or external to the industrial site) is quite similar to an exchange of materials, energy, or by-products among different industrial facilities.

4.3. Definition of the Database Structure

The database is organized as follows: each row of the database corresponds to a low-temperature heat recovery case study (for each bibliographic source there can be multiple case studies).

The information collected is organized into seven categories:

1. Identification: identification of each single case study within the database. For each source, there can be multiple WHR case collected.
2. Source: reference to the document in which the case study is described.
3. Company: this category geographically locates the heat recovery project and includes the industrial sectors and subsectors involved.
4. Waste heat: includes all the information necessary to characterize the available waste heat source.
5. Recovery process: includes information about the type of recovery intervention performed and the technology used, following the structure defined by the *Technology database*.
6. Waste heat utilization: contains information describing how the recovered waste heat is used.
7. Other information: refers to other technical, economic, and general information not recorded in the previous fields that may help the user to evaluate and understand the WHR application.

Table 6 shows all the database fields and their detailed description.

Table 6. *Case study database*'s fields and description.

Category	Field	Description
Identification	Document ID	A unique progressive number assigned to the source
	Project ID	A unique progressive number assigned to the heat recovery project
Source	Authors	Name of the authors
	Year	Year of publication
	Source type	Scientific article, project report, etc.
	Journal	In the case of a scientific article, it indicates the publication journal
	Link	Source reference (e.g., DOI [1], URL [2], etc.)
Company	Geographical reference	The country in which the WHR [3] project is implemented
	Sectors	Sectors involved in the production of waste heat
	Sub-sectors	In the case of a more detailed identification of the sector
	Other sectors involved	Sectors involved in the use of waste heat (if different from the previous one)
Waste Heat	Process	Description of the process that produces waste heat
	Vector	Source from which the heat is recovered (e.g., hot water, exhaust gas)
	Quantity	Flow, thermal power or heat produced
	Availability	Frequency of waste heat production (e.g., hour per year)
	Temperature	Waste heat temperature
	Flow rate	Waste heat flow rate

Table 6. *Cont.*

Category	Field	Description
Recovery process	Recovery type	Indicates the type of heat recovery (e.g., waste heat to electricity, waste heat to cooling, etc.)
	Technology	WHR[3] technology used (e.g., ORC [4], heat pumps, etc.)
	Technology description	Additional information on the technology used
	State of maturity	It represents the state of maturity of the technology (e.g., consolidated, emerging, etc.)
	Vector fluid	Vector fluid used in the recovery process (e.g., water, oil, etc.)
	Vector fluid quantity	Vector fluid flow rate
Waste heat utilization	Receiving process	The process that receives the recovered energy
	Internal/External	Specifies the nature of the heat exchange, which can be internal or external to the process
	Quantity	Flow, thermal power or heat produced
	Use	Frequency of waste heat utilization (e.g., hour per year)
	Temperature	The temperature at which the "recovered" energy carrier is used
Other information	Barriers	Main barriers encountered in project implementation
	Identified solutions	If during the implementation of the project solutions were evaluated to overcome barriers
	Notes	Other relevant information about the WHR[3] project

[1] Digital object identifier (DOI); [2] Uniform resource locator (URL); [3] Waste heat recovery (WHR); [4] Organic Rankine cycle (ORC).

Finally, all the information collected from each source was implemented into the database.

4.4. Population of the Database

Documents selected during the analysis of scientific literature (130 documents) were further screened to identify studies reporting most of the information required by the database fields previously described (Table 6).

There were 81 sources used for the *Case study database* population, corresponding to 108 case studies of waste heat recovery. Figure 6a shows that the sources were primarily scientific articles, mainly published from 2017 to 2019 (Figure 6b). Regarding the individual WHR cases, Figure 7a shows the industrial sector breakdown. It is noted that most of the studies concerned the food and the textile sectors.

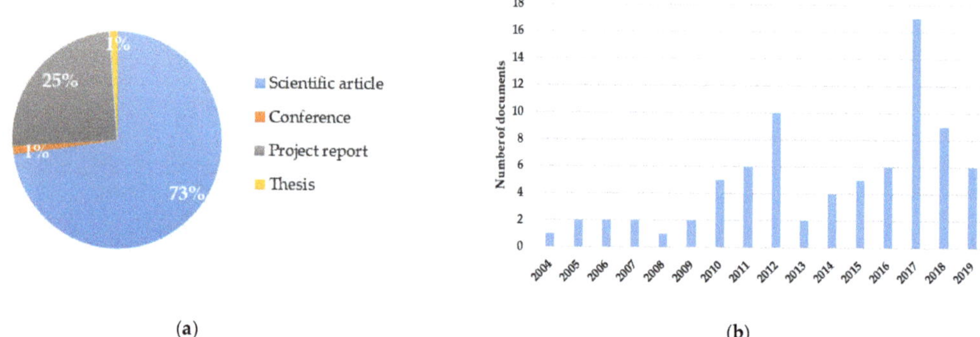

Figure 6. Results of the *Case study database*'s preliminary population: (**a**) Type of source analyzed; (**b**) Publication year of different sources.

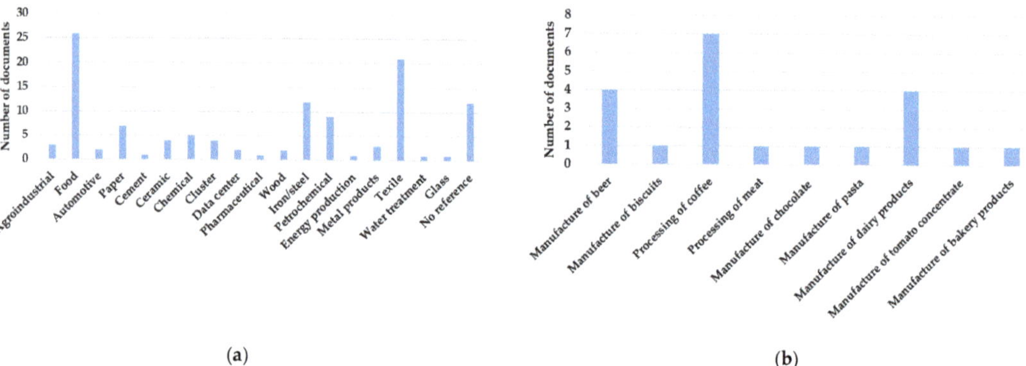

Figure 7. Results of the *Case study database*'s preliminary population: (**a**) Industrial sectors involved in recovery cases; (**b**) Food subsectors involved in recovery cases.

Concerning the food sector, which was the most numerous and detailed sector, Figure 7b also shows the relative sub-sectors, including coffee, dairy, and beer production.

It is interesting to note that the majority of case studies within the database are related to internal process heat recovery. However, this aspect may have been emphasized by the type of research carried out, as most of the articles focusing on external heat recovery (e.g., industrial symbiosis) do not report detailed information about the waste heat flows and were therefore excluded during the database population.

5. Use of the Databases

The two databases described represent two key elements of a simulation tool supporting the identification and evaluation of suitable waste heat recovery opportunities in industrial facilities.

The choice of creating two distinct databases is confirmed to be positive for enhancing the modularity of the final decision support tool, allowing us to obtain two related but unconstrained tools to be used in distinct phases of the decision-making process.

Furthermore, this distinction has certainly made it possible to obtain advantages from the two databases' population, focusing on different aspects and collecting more information maintaining a simple structure.

Although the two tools are distinct, they can be jointly used, as it is possible to integrate the information found from one of the two databases by consulting the other.

In order to provide a practical example of the integration of the two databases, two possible processes for the joint use of the two tools have been identified:

(a) User A consults the *Case study database* searching for heat recovery applications in a specific sector or sub-sector;

(b) User B consults the *Technology database* looking for a technology suitable for the available waste heat characteristics.

Figure 8 shows a graphical representation and description of the proposed example.

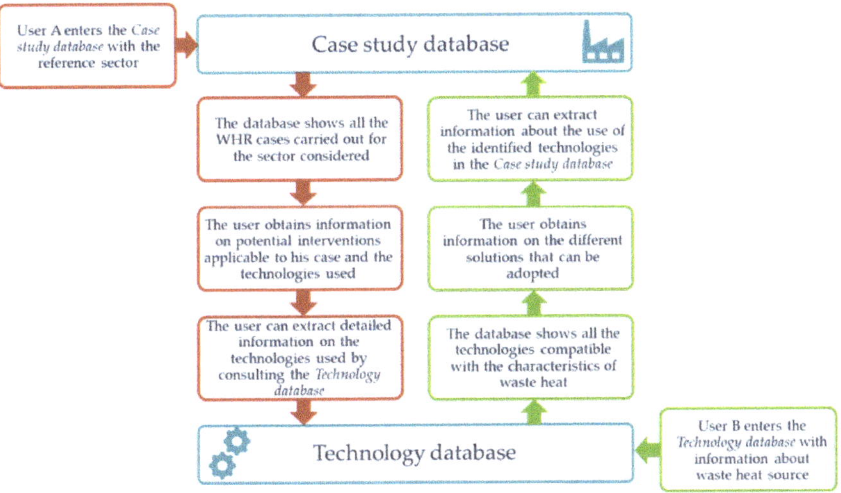

Figure 8. Two different integrated uses of the databases.

User A's query represents the common example of a user looking for opportunities to improve efficiency through waste heat recovery as a company with significant thermal consumption. He wants to get an idea of the recovery interventions already applied in his sector (sub-sector or process), and analyze whether there are also margins for implementation in his company. However, User B's query represents the example of a user who has already identified a waste heat to use and who therefore wants to obtain more information about the technologies that can be implemented to recover it. Furthermore, User B may already be aware of the technologies that can be implemented (for example, downstream of a preliminary assessment), but is looking for more detailed information.

It is essential to highlight that those proposed are only two of the databases' possible uses. For both tools, the piece of information used to access the database could be different, even if the process of retrieving additional information can follow a similar logic to the ones proposed in Figure 8.

To give a more practical example, we imagine an Energy Manager [40] from a dairy company searching for WHR projects implemented in this sub-sector. Entering "Manufacture of dairy products" in the sub-sectors field of the *Case study database*, he would find four case studies. Two of these refer to waste heat recovery through the application of heat pumps. To obtain more information about this technology, the user could consult the *Technology database* entering "Heat pump" in the technology field. The search would show 21 results, from which the user would have more detailed information about heat pumps and he could then select those compatible with the characteristics of the waste heat flow available from the dairy company.

Table 7 provides some extracts of the results obtained using the search criteria described. Due to confidentiality issues, the table's data do not provide any information about the companies and the technology provider involved.

Table 7. Extract of the results obtained using the search criteria specified in the example of the use of databases.

Case Study Database						
Source Type	**Sub-Sectors ***	**Process**	**Receiving Process**	**Technology**	**Quantity**	**Temperature**
Project report	Manufacture of dairy products	Process water cooling	Production of hot water for heating of a nearby greenhouse	Heat pump	7.2 MW	73 °C
Project report	Manufacture of dairy products	Recovery of waste heat from the refrigeration system condenser	Preheating of the air to produce powdered milk	Heat pump	1.25 MW	80 °C
Technology database						
Source type	**Recovery type**	**Technology ***	**State of maturity**	**Vector fluid**	**Output power**	**Output temperature**
Catalog	Waste Heat to Heat	Heat pump	Consolidated	Water	0.6–2.0 MW	80
Catalog	Waste Heat to Heat	Heat pump	Consolidated	Water	0.6–10.0 MW	70–90

* The fields marked with an asterisk are the fields used to enter the two databases in this specific example.

To maintain search effectiveness, in addition to those mentioned, the user can consult the database using more specific fields, such as the type of fluid, the power range, the state of maturity of the technology, the process, the receiving process, etc. By inserting more detailed search criteria, the problems of using the database in terms of an excessive number of results can be limited, even in the case of an extensive database.

The information contained in the two databases allows the energy attributes for any WHR system to be demonstrated. Referring to the technology found in Table 7, Figure 9 reports an example made by a Sankey diagram of a heat pump. In the diagram, the electricity is shown in green, while the thermal flows are shown in a color scale, from bright red for high temperatures to blue for lower temperatures.

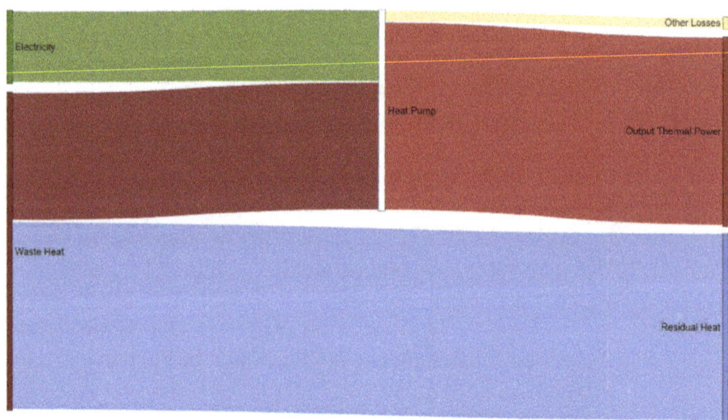

Figure 9. Example of a Sankey Diagram for a heat pump.

6. Discussion and Conclusions

This article presents the main phases that led to the development of two databases aimed at supporting companies in increasing the diffusion of low-temperature heat recovery interventions in the industrial sector, which, despite the high potential, are not sufficiently applied.

The first database is focused on available state-of-the-art and advanced technologies, while the second database reports several WHR case studies that have already been implemented.

Compared to the other tools related to energy efficiency, the two databases proposed in this article are the only ones entirely focused on low-temperature waste heat recovery. Although the literature about heat recovery is extensive and presents numerous remarkable contributions, these previous attempts do not result in tools that are effectively usable by companies. Therefore, they do not represent a valid aid in overcoming those gaps in the diffusion of recovery interventions, which we have found to be mainly due to non-technological barriers.

Compared to the existing tools and databases, one of the main advantages of the proposed databases is that they contain all the information necessary to analyze the heat recovery process both from a qualitative and a quantitative perspective. For these reasons, the databases represent two fundamental elements of a complete information system that will be developed into a broader research project, aimed at supporting companies in identifying and conducting the first evaluation of heat recovery opportunities and overcoming non-technical barriers to the diffusion of energy efficiency measures.

In order to favor the modularity of the final tool, two distinct databases were created to be applied in different phases of the implementation of a heat recovery project. A common procedure was followed to develop both tools, starting from an in-depth analysis of the literature, which provided the basis for identifying an appropriate database structure and for providing the inputs for its preliminary population.

The *Case study database* contains 31 fields divided into seven categories. It allows for the collection of information relating to the industrial sector and subsector in which the case study is implemented, the characteristics of waste heat, the technology used, and information about the reuse of the heat recovered. The *Technology database* contains 28 fields divided into five categories. It permits collecting information relating to the technology provider, technology description and state of maturity, and information about the technical characteristics and range of usability for the technologies collected.

The databases' structure was preliminarily validated thanks to the interaction with important stakeholders and possible users of the tools, such as companies operating in the industrial sector and some WHR technology suppliers operating in the Italian and European markets. This interaction provided important feedback about the two proposed tools' usability and helped to define a more appropriate database structure. Therefore, this allowed for the consolidation of the data within the *Technology database,* regarding several aspects of both well-established and emerging technologies, such as the typical operating temperature and the power range, the energy performance, and the economic parameters (i.e., specific costs, operating and maintenance costs, lifetime, etc.), and the main barriers to the technology implementation. The interaction with technology manufacturers also allowed us to increase the size of the *Case study database,* by providing data regarding real case studies on WHR from industrial wastes.

Meetings with industries which use different types of processes were also organized to verify and consolidate data within the *Case study database,* particularly regarding types and characteristics of waste heat sources generated by the production processes and the types of WHR solutions usually implemented, depending on the industrial sector under consideration.

Furthermore, possible integrated use of databases and a practical example of their utilization were proposed. This application showed how much information on heat recovery interventions and implementable technologies could be obtained in a typical situation with little input information (such as industrial sector or waste heat characteristics).

It should be noted that the two databases were only preliminary populated, and their content will be substantially increased, thanks to the analysis of energy audits and visits to different companies involved, during the following years of the project.

Advanced data analysis techniques, such as machine learning approaches, could be used when the databases are sufficiently populated. Possible applications are, for example, database schema matching [41,42], in order to improve the integration of the two databases, or using the available data to model the relationships between the various parameters such as output power, waste heat temperature, and type of fluid for the different technologies [43].

The following steps will involve the IT platform development in which the two databases will represent key tools. Together with other modules, such as the pinch analysis, the evaluation of external uses, and the development of specific models of the single technologies, this will guide the user in all the different phases of implementing a WHR intervention.

Author Contributions: All authors contributed to the idea and the design of the proposed tools; D.D. was responsible for the *Technology database* development; M.B., L.G. and P.E.L. were responsible for the *Case study database* development; M.B., D.D. and L.G. prepared the original draft; V.I. and A.S. were responsible for the activities carried out by University of Rome Tor Vergata and contributed to the review and editing. Project manager: M.B. All authors have read and agreed to the published version of the manuscript.

Funding: The research activity has been funded by the Project "Energy efficiency of industrial products and processes", Research programme "Piano Triennale della Ricerca del Sistema Elettrico Nazionale 2019–2021" funded by the Italian Ministry of Economic Development.

Institutional Review Board Statement: Not applicable.

Informed Consent Statement: Not applicable.

Data Availability Statement: Not applicable.

Conflicts of Interest: The authors declare no conflict of interest. The funders had no role in the design of the study; in the collection, analyses, or interpretation of data; in the writing of the manuscript, or in the decision to publish the results.

References

1. Moran, E.F.; Lopez, M.C.; Moore, N.; Müller, N.; Hyndman, D.W. Sustainable Hydropower in the 21st Century. *Proc. Natl. Acad. Sci. USA* **2018**, *115*, 11891–11898. [CrossRef]
2. Okot, D.K. Review of Small Hydropower Technology. *Renew. Sustain. Energy Rev.* **2013**, *26*, 515–520. [CrossRef]
3. Kumar, Y.; Ringenberg, J.; Depuru, S.S.; Devabhaktuni, V.K.; Lee, J.W.; Nikolaidis, E.; Andersen, B.; Afjeh, A. Wind Energy: Trends and Enabling Technologies. *Renew. Sustain. Energy Rev.* **2016**, *53*, 209–224. [CrossRef]
4. Tiwari, R.; Babu, N.R. Recent Developments of Control Strategies for Wind Energy Conversion System. *Renew. Sustain. Energy Rev.* **2016**, *66*, 268–285. [CrossRef]
5. Santolamazza, A.; Dadi, D.; Introna, V. A Data-Mining Approach for Wind Turbine Fault Detection Based on SCADA Data Analysis Using Artificial Neural Networks. *Energies* **2021**, *14*, 1845. [CrossRef]
6. Kabir, E.; Kumar, P.; Kumar, S.; Adelodun, A.A.; Kim, K.-H. Solar Energy: Potential and Future Prospects. *Renew. Sustain. Energy Rev.* **2018**, *82*, 894–900. [CrossRef]
7. Gong, J.; Li, C.; Wasielewski, M.R. Advances in Solar Energy Conversion. *Chem. Soc. Rev.* **2019**, *48*, 1862–1864. [CrossRef] [PubMed]
8. Sundaram, M.M.; Appadoo, D. Traditional Salt-in-Water Electrolyte vs. Water-in-Salt Electrolyte with Binary Metal Oxide for Symmetric Supercapacitors: Capacitive vs. Faradaic. *Dalton Trans.* **2020**, *49*, 11743–11755. [CrossRef] [PubMed]
9. Minakshi, M.; Mitchell, D.R.G.; Jones, R.T.; Pramanik, N.C.; Jean-Fulcrand, A.; Garnweitner, G. A Hybrid Electrochemical Energy Storage Device Using Sustainable Electrode Materials. *ChemistrySelect* **2020**, *5*, 1597–1606. [CrossRef]
10. Sharma, P.; Minakshi, M.; Whale, J.; Jean-Fulcrand, A.; Garnweitner, G. Effect of the Anionic Counterpart: Molybdate vs. Tungstate in Energy Storage for Pseudo-Capacitor Applications. *Nanomaterials* **2021**, *11*, 580. [CrossRef] [PubMed]
11. Braff, W.A.; Mueller, J.M.; Trancik, J.E. Value of Storage Technologies for Wind and Solar Energy. *Nat. Clim. Chang.* **2016**, *6*, 964–969. [CrossRef]
12. Firth, A.; Zhang, B.; Yang, A. Quantification of Global Waste Heat and Its Environmental Effects. *Appl. Energy* **2019**, *235*, 1314–1334. [CrossRef]
13. Brunner, C.; Muster, B.; Heig, E.; Schweiger, H.; Vannoni, C. Einstein—Expert System for an Intelligent Supply of Thermal Energy in Industry—Audit Methodology and Software Tool. *Chem. Eng. Trans.* **2010**, *21*, 685–690. [CrossRef]

14. EINSTEIN II. Available online: https://www.einstein-energy.net/index.php/about-einstein/fundedprojects/einstein-ii (accessed on 23 March 2021).

15. Glatzl, W.; Brunner, C.; Fluch, J. GREENFOODS—Energy Efficiency in the Food and Beverage Industry. In Proceedings of the Eceee Summer Study on Energy Efficiency, Toulon/Hyères, France, 1–6 June 2015.

16. SEEnergies Project. Available online: https://www.seenergies.eu (accessed on 23 March 2021).

17. Forman, C.; Muritala, I.K.; Pardemann, R.; Meyer, B. Estimating the Global Waste Heat Potential. *Renew. Sustain. Energy Rev.* **2016**, *57*, 1568–1579. [CrossRef]

18. Xu, Z.Y.; Wang, R.Z.; Yang, C. Perspectives for Low-Temperature Waste Heat Recovery. *Energy* **2019**, *176*, 1037–1043. [CrossRef]

19. Jouhara, H.; Khordehgah, N.; Almahmoud, S.; Delpech, B.; Chauhan, A.; Tassou, S.A. Waste Heat Recovery Technologies and Applications. *Therm. Sci. Eng. Prog.* **2018**, *6*, 268–289. [CrossRef]

20. Martin, M.; Harris, S. Prospecting the Sustainability Implications of an Emerging Industrial Symbiosis Network. *Resour. Conserv. Recycl.* **2018**, *138*, 246–256. [CrossRef]

21. Dénarié, A.; Fattori, F.; Spirito, G.; Macchi, S.; Cirillo, V.F.; Motta, M.; Persson, U. Assessment of Waste and Renewable Heat Recovery in DH through GIS Mapping: The National Potential in Italy. *Smart Energy* **2021**, *1*, 100008. [CrossRef]

22. Wheatcroft, E.; Wynn, H.; Lygnerud, K.; Bonvicini, G.; Leonte, D. The Role of Low Temperature Waste Heat Recovery in Achieving 2050 Goals: A Policy Positioning Paper. *Energies* **2020**, *13*, 2107. [CrossRef]

23. Department of Energy (DOE). Waste Heat Recovery Systems. Chapter 6: Innovating Clean Energy Technologies in Advanced Manufacturing-Technology Assessments. In *Quadrennial Technology Review 2015-An Assessment of Energy Technologies and Research Opportunities*; DOE: Washington, DC, USA, 2015.

24. Haddad, C.; Périlhon, C.; Danlos, A.; François, M.-X.; Descombes, G. Some Efficient Solutions to Recover Low and Medium Waste Heat: Competitiveness of the Thermoacoustic Technology. *Energy Procedia* **2014**, *50*, 1056–1069. [CrossRef]

25. Simeone, A. A Decision Support System for Waste Heat Recovery in Manufacturing. *CIRP Ann.* **2016**, *65*, 21–24. [CrossRef]

26. Ling-Chin, J.; Bao, H.; Ma, Z.; Taylor, W.; Paul Roskilly, A. State-of-the-Art Technologies on Low-Grade Heat Recovery and Utilization in Industry. In *Energy Conversion—Current Technologies and Future Trends*; Al-Bahadly, I.H., Ed.; IntechOpen: Londone, UK, 2019; ISBN 978-1-78984-904-2.

27. Kovala, T.; Wallin, F.; Hallin, A. Factors Influencing Industrial Excess Heat Collaborations. *Energy Procedia* **2016**, *88*, 595–599. [CrossRef]

28. Grönkvist, S.; Sandberg, P. Driving Forces and Obstacles with Regard to Co-Operation between Municipal Energy Companies and Process Industries in Sweden. *Energy Policy* **2006**, *34*, 1508–1519. [CrossRef]

29. Broberg Viklund, S.; Karlsson, M. Industrial Excess Heat Use: Systems Analysis and CO2 Emissions Reduction. *Appl. Energy* **2015**, *152*, 189–197. [CrossRef]

30. The EU-MERCI Project. Available online: http://www.eumerci.eu (accessed on 23 March 2021).

31. Maggiore, S.; Realini, A.; Zagano, C. 'Good Practices' to Improve Energy Efficiency in the Industrial Sector. *Int. J. Energy Prod. Manag.* **2020**, *5*, 259–271. [CrossRef]

32. IAC: Industrial Assessment Centers. Available online: https://iac.university (accessed on 23 March 2021).

33. MAESTRI Project. Available online: http://maestri-spire.eu/ (accessed on 23 March 2021).

34. SHIP Plants. Available online: http://ship-plants.info (accessed on 23 March 2021).

35. AEE—Institut Für Nachhaltige Technologien—"Matrix of Industrial Process Indicators". Available online: http://wiki.zero-emissions.at/index.php?title=Main_Page (accessed on 23 March 2021).

36. Brückner, S.; Liu, S.; Miró, L.; Radspieler, M.; Cabeza, L.F.; Lävemann, E. Industrial Waste Heat Recovery Technologies: An Economic Analysis of Heat Transformation Technologies. *Appl. Energy* **2015**, *151*, 157–167. [CrossRef]

37. Papapetrou, M.; Kosmadakis, G.; Cipollina, A.; La Commare, U.; Micale, G. Industrial Waste Heat: Estimation of the Technically Available Resource in the EU per Industrial Sector, Temperature Level and Country. *Appl. Therm. Eng.* **2018**, *138*, 207–216. [CrossRef]

38. Dadi, D.; Introna, V.; Cesarotti, V.; Benedetti, M. Design and Development of a Database to Identify and Evaluate Waste Heat Recovery Opportunities. In Proceedings of the Summer School Francesco Turco, Bergamo, Italy, 9–11 September 2020.

39. Johnson, I.; Choate, W.T.; Davidson, A. *Waste Heat Recovery: Technology and Opportunities in U.S. Industry*; BCS, Inc.: Laurel, MD, USA; Washington, DC, USA, 2008; p. 112.

40. L. 10/1991 Norme per L'attuazione del Piano Energetico Nazionale in Materia di uso Razionale Dell'energia, di Risparmio Energetico e di Sviluppo Delle Fonti Rinnovabili di Energia. 1991. Available online: https://www.gazzettaufficiale.it/eli/id/1991/01/16/091G0015/sg (accessed on 23 March 2021).

41. Berlin, J.; Motro, A. Database Schema Matching Using Machine Learning with Feature Selection. *Semin. Contrib. Inf. Syst. Eng.* **2013**, *15*, 315–329.

42. Rahm, E.; Bernstein, P.A. A Survey of Approaches to Automatic Schema Matching. *VLDB J.* **2001**, *10*, 334–350. [CrossRef]

43. Benedetti, M.; Dadi, D.; Introna, V.; Santolamazza, A. Proposal of a Methodology for the Preliminary Assessment of Low Temperature Heat Recovery Opportunities for Industrial Applications. In Proceedings of the International Conference on Applied Energy 2020 (ICAE 2020), Bangkok/Virtual, Thailand, 1–10 December 2020.

 *sustainability*

Article

Cogeneration Supporting the Energy Transition in the Italian Ceramic Tile Industry

Lisa Branchini [1,*], Maria Chiara Bignozzi [2], Benedetta Ferrari [2], Barbara Mazzanti [3], Saverio Ottaviano [1], Marcello Salvio [4], Claudia Toro [4], Fabrizio Martini [4] and Andrea Canetti [5]

[1] Department of Industrial Engineering, University of Bologna, Viale del Risorgimento 2, 40136 Bologna, Italy; saverio.ottaviano2@unibo.it

[2] Department of Civil, Chemical, Environmental and Materials Engineering, University of Bologna, Via Terracini 28, 40131 Bologna, Italy; maria.bignozzi@unibo.it (M.C.B.); benedetta.ferrari13@unibo.it (B.F.)

[3] Centro Ceramico, Via Tommaso Martelli, 26, 40138 Bologna, Italy; mazzanti@centroceramico.it

[4] DUEE-SPS-ESE Laboratory, Italian National Agency for New Technologies, Energy and Sustainable Economic Development (ENEA), Lungotevere Thaon di Revel 76, 00196 Rome, Italy; marcello.salvio@enea.it (M.S.); claudia.toro@enea.it (C.T.); fabrizio.martini@enea.it (F.M.)

[5] Confindustria Ceramica, Viale Monte Santo, 40, 41049 Sassuolo, Italy; acanetti@confindustriaceramica.it

* Correspondence: lisa.branchini2@unibo.it; Tel.: +39-0512093314

Abstract: Ceramic tile production is an industrial process where energy efficiency management is crucial, given the high amount of energy (electrical and thermal) required by the production cycle. This study presents the preliminary results of a research project aimed at defining the benefits of using combined heat and power (CHP) systems in the ceramic sector. Data collected from ten CHP installations allowed us to outline the average characteristics of prime movers, and to quantify the contribution of CHP thermal energy supporting the dryer process. The electric size of the installed CHP units resulted in being between 3.4 MW and 4.9 MW, with an average value of 4 MW. Data revealed that when the goal is to maximize the generation of electricity for self-consumption, internal combustion engines are the preferred choice due to higher conversion efficiency. In contrast, gas turbines allowed us to minimize the consumption of natural gas input to the spray dryer. Indeed, the fraction of the dryer thermal demand (between 600–950 kcal/kgH$_2$O), covered by CHP discharged heat, is strictly dependent on the type of prime mover installed: lower values, in the range of 30–45%, are characteristic of combustion engines, whereas the use of gas turbines can contribute up to 77% of the process's total consumption.

Keywords: cogeneration; energy efficiency; ceramic industry; spray dryer; energy analysis; process thermal consumption; gas turbine; internal combustion engine

Citation: Branchini, L.; Bignozzi, M.C.; Ferrari, B.; Mazzanti, B.; Ottaviano, S.; Salvio, M.; Toro, C.; Martini, F.; Canetti, A. Cogeneration Supporting the Energy Transition in the Italian Ceramic Tile Industry. *Sustainability* **2021**, *13*, 4006. https://doi.org/10.3390/su13074006

Academic Editor: Gerardo Maria Mauro

Received: 11 March 2021
Accepted: 31 March 2021
Published: 3 April 2021

Publisher's Note: MDPI stays neutral with regard to jurisdictional claims in published maps and institutional affiliations.

1. Introduction

The topic of energy efficiency in industry is rising to the top of the agendas of the European Union (EU) and Member States, primarily for environmental (need to reduce greenhouse gas emissions) and economic (i.e., unstable energy prices and reliability of supply) arguments [1,2].

The Energy Efficiency Directive 2012/27/EU (EED) is a solid cornerstone of Europe's energy legislation. It includes a balanced set of binding measures planned to help the EU reach its 20% energy efficiency target by 2020. The EED establishes a common framework of measures for the promotion of energy efficiency (EE) to ensure the achievement of the European targets and to pave the way for further EE improvements beyond 2020. The Italian Government transposed the EED in 2014 (by issuing the Legislative Decree no. 102/2014, recently updated by Legislative Decree no. 73/2020), also extending the obligation to a specific group of energy-intensive enterprises (mostly small and medium enterprises, SME). The ENEA (Italian National Agency for New Technologies, Energy, and the Sustainable Economic Development) was appointed to manage the obligation of EED Article 8 [3], which is dedicated to energy audits, a tool used to assess the existing energy consumption and identify the whole range of opportunities to save energy.

In the EED, energy audit (EA) is defined as a systematic procedure aimed at obtaining adequate knowledge on the existing energy consumption profile of a building or group of buildings, an industrial or commercial operation, or installation for private or public service, identifying and quantifying cost-effective energy savings opportunities, and reporting the findings.

According to Article 8 of Legislative Decree 102/14, two categories of companies, namely large enterprises and energy-intensive enterprises, have been targeted as obliged to carry out energy audits on their sites, first by the 5 December 2015, and then at least every four years.

Obliged enterprises that do not carry out an energy audit observing Annex II of the EED within the above-mentioned deadlines will be subject to administrative monetary penalties. According to Article 8 of the Italian Legislative Decree 102/2014 implementing the Energy Efficiency Directive, as of 31 December 2019, ENEA received 11,172 energy audits of production sites, related to 6434 companies [4].

Over 53% of the audits were carried out on sites related to the manufacturing sector and over 14% in trade. A total of 70% of the audits collected by ENEA are equipped with specific monitoring of energy consumption.

Despite relevant efforts having been deployed in terms of both innovation technologies and regulatory frameworks in enabling it, the full potential for energy efficiency in the industrial sector remains significant.

According to [5], in 2011, industry was the largest heat-consuming sector (79 EJ), accounting for 46% of the world total thermal energy demand. Based on recent IEA Outlooks [6,7], however, the industrial heat makes up two-thirds of industrial energy demand and 20% of global energy consumption.

In Italy, in 2018 [8], about 2234 ktoe of heat was consumed by industry, representing 53% of the national heat demand, 7% of the total national energy need, and 17.5% of the industry energy need. Most of the required heat comes from the direct combustion of fuels, with natural gas as the main supplier followed by petroleum and coal products. The electricity consumption of the Italian industrial sector, according to [9], with reference to 2018, was equal to 126.4 10^9 kWh, representing about 42% of the national electricity request.

Within the framework of the described scenario, the reduction in energy consumption in the industrial sector, together with the increment of the efficiency of energy generation technologies, are fundamental aspects to meet the EU targets. The industrial sector offers tremendous opportunity for low-cost energy savings and carbon reductions through energy efficiency improvements.

In this context, cogeneration, or combined heat and power (CHP) (i.e., the combined generation of electricity and useful heat from a single primary energy source), still plays a significant role while maintaining long-term prospects in the global energy markets, primarily due to its numerous operational, environmental, and economic benefits. Since the 1970s, cogeneration has been used to improve the efficiency of production systems, both in the industrial and civil fields. In Italy, industry is the sector that invested the most in cogeneration technologies in the past decades. The higher value of the electricity price paid by Italian industries compared to other EU members was one of the main drivers toward the diffusion of CHP. Indeed, the sum of investments in energy efficiency in 2018 in Italy was about 7.1 billion €, of which about 7% (about 480 M€) was in cogeneration technologies [10]. The industrial sector has contributed for 443 million € of investments in CHP technologies.

According to the latest annual report on cogeneration [11], the total number of CHP units installed in Italy is equal to 1737, more than 89% of which are represented by internal combustion engine (ICE) units. The overall installed generation capacity is equal to 13,233 MW. The annual gross electricity generation with CHP units equals 58,722 GWh/y, while useful heat generation reached 36,076 GWh/y. Natural gas is the main source, with about 119,000 GWh/y of primary energy supplied.

Application of CHP technologies in the industrial sector is preferred in manufacturing processes that require a significant amount of electricity and heat simultaneously, throughout the whole year. In this respect, the ceramic tile production is one among the industrial processes where energy efficiency management is crucial, given the high energy demand necessary for the production cycle (covered, at the present, for about 70% by natural gas), and the incidence of the energy item on the final production cost, close to 20%.

In 2016, Confindustria Ceramica, the association of Italian ceramics, conducted research on thermal efficiency strategies developed to optimize the energy consumption of the ceramic tiles industry [12]. The research analyzed a sample of 64 production sites covering 59.6% of national production. A first energy saving solution, started in the early 1980s, was the recovery of the indirect cooling air of kilns to the dryers. In 2016, this strategy concerned 43.2% of vertical dryers and 35.2% of horizontal dryers. The second and most effective energy efficiency strategy was the introduction of the CHP units in the 90s, and their diffusion had another strong impulse around 2008 (see Figure 1), thanks to the establishment in Italy of the TEE ("Titoli di Efficienza Energetica"). The TEE is a mechanism introduced by D.M. 24 April 2001 to incentivize the implementation of energy efficiency interventions that comply with the national energy savings targets (2001/77/CE). The most adopted configuration provides for the spray dryer as the thermal user for the CHP plant, while the prime movers are gas turbines and internal combustion engines. The study reported in [12] also highlighted that most parts of the CHP units with gas turbine technology was closer to their end life (17.4 years compared to a hypothetical end life of 24 years) than CHP units with an internal combustion engine (eight years compared to a hypothetical end life of 20 years). This means that this latter technology has generally been installed more frequently since 2008.

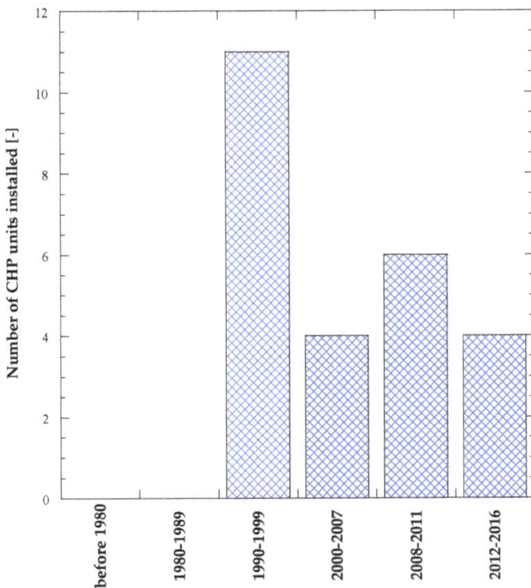

Figure 1. Combined heat and power (CHP) units installed in the period 1980–2016 [12].

In 2019, the ceramic tile industries located in Italy comprised 135 [13], of which 54 were equipped with a spray dryer. Within the latter, 28 were provided with a CHP unit, helping to satisfy both the electrical and thermal request of the production process.

The literature on cogeneration in industrial applications is exhaustive: most relevant examples can be found in [14–25]. In detail, performance assessment and optimal scheduling of a CHP power plant for industrial applications can be found in [14–18]. Description and energy analysis of specific applications of CHP to industrial processes have been reported in [20–24] for the food processing industry, in [21] for cement production, in [25] for the textile industry, and finally in [19] for the chemical and pulp mill industry.

In contrast, studies dealing with the specific application of cogeneration in ceramic tiles factories, according to the authors' knowledge, are limited. Caglayana et al. [26–28] presented an energy, exergy, and sustainability analysis of a CHP gas turbine applied in the ceramic sector. The main results of their studies showed that the utilization of the gas turbine (GT) unit could provide 0.1115 m^3/s and 0.0732 m^3/s of natural gas savings for the ground and wall tile dryers, respectively. Moreover, their results highlight that the most efficient components are the air compressor and combustion chamber, while the minimum energy efficiencies were obtained for the wall tile dryer (8%) and ground tile dryer (8%) components. Yoru et al. [29] presented an energy and exergy analysis on a CHP system installed into a ceramic plant consisting of three GT units. Results claim that the exergy efficiency was inversely proportional to the ambient temperature: the rise of ambient temperature at the compressor inlet adversely affected the efficiency of the system. The energy and exergy efficiencies of the cogeneration system were calculated as 82% and 35%, respectively.

More strategies on how the EU ceramic sector is following energy efficiency policies and environmental concerns can be found in [30–36].

In particular, different technological options to reach the EU 2020 and 2050 greenhouse gas (GHG) emissions objectives were compared in [34] using the LCA methodology on 25 different technological scenarios of the life cycle of porcelain stone tiles. The GHG emissions' objectives, considering only the ceramic tile production stage, can be achieved by modifying the product design (removal of the glaze and reduction of the thickness of the ceramic body) or by electrifying 50% of the thermal processes through renewable sources.

A more efficient and sustainable production appears to be the only option for a long-term prospective of manufacturing industries. Efforts required by the traditional industry should be directed toward implementing measures that concurrently guarantee a global energy savings and a reduction in the environmental impacts.

In the aforementioned context, ENEA launched a two-year research project in collaboration with the University of Bologna, Centro Ceramico, and Confindustria Ceramica with the aim of delineating the current status of cogeneration in the Italian ceramic tile sector. The final purpose of the project is to define the energetic and environmental benefits associated with the application of CHP systems to the ceramics industry. This paper summarizes the results of the first-year research activities, which focused on outlining the average characteristics of the prime movers installed, and quantifying the contribution of CHP thermal energy, supporting the spray dryer section. Starting from the analysis of the energy audits of ceramic companies collected by ENEA, a detailed assessment of the CHP units' specific characteristics was realized by integrating these data with information directly requested from ceramic companies.

The study's outcomes provide for an overview of the current situation, and can be used by industrial developers as guidelines for the selection and design of CHP units as well as by law-making bodies as a reference for energy efficiency programs including incentives policy.

The rest of this manuscript is organized as follows. In Section 2, a synthetic description of the ceramic tile production process is presented, and emphasis is given to the detailed explanation of the spray dryer technology. Thermal integration with the CHP units, contributing to satisfy the heat consumption of the dryer unit, is presented with schematic layouts of the integrated system. The research method, data collection, and energetic performance indexes, ad hoc defined for the purpose of this analysis, are outlined in Section 3. Section 4 presents and discusses the main results of the study. Finally, in Section 5, the conclusions of the study are highlighted along with the future steps of the project.

2. Description of the Ceramic Tile Process

Ceramic tile production plants have been totally revolutionized to improve productivity and decrease energy consumption as a reaction to the energy crisis of 1974 [37].

Ceramic tiles are the result of a process that is strongly influenced by the raw materials and firing temperature: as a function of these two parameters, different types of tiles can be obtained (porcelain tile, monoporosa, single-firing tile, double-firing tile, etc.).

The production process (Figure 2) generally starts with the storage and the preparation phase of the raw materials, consisting of two main operations: wet grinding and water content adjustment [38]. Wet grinding technology, which uses a continuous or discontinuous drum mill, can reduce particle size, dust pollution, and provides a good homogenization [39]. Subsequently, the aqueous suspension ("slip", with a water content of ≅30%) obtained from the wet grinding is dried and transformed into spherical granules by way of the operation called spray drying.

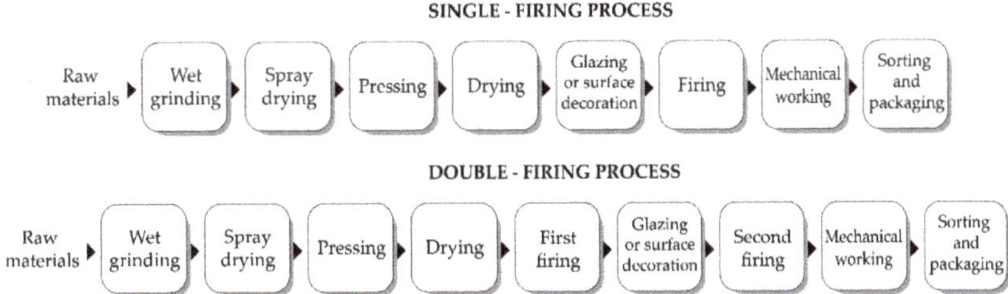

Figure 2. Flow chart of the single and double production process of ceramic tiles.

Spray-dried granules (still containing ≅5–6% of water) are then shaped by discontinuous or continuous pressing in order to obtain the tiles, which are further dried to remove the remaining water through a process of surface evaporation and interstitial diffusion, which takes place homogeneously in the vertical or horizontal rapid dryers. After the glazing and decoration phases, currently mainly performed by digital printing, firing takes place in roller kilns 70 ÷ 100 m long. In kilns, the tiles are exposed to increasing temperature until they reach the maximum temperature zone around 1000–1200 °C, then they are quickly cooled. After cooling, tiles can be treated by several mechanical processes such as cutting, grinding, lapping, and polishing, after which they are sorted and packed.

The production process above described is a single firing process and is mainly used for the production of monoporosa and porcelain tiles.

A double firing process was also adopted in the past. In this case, glazing and decoration occurred after a first firing step at about 1000–1100 °C, which is needed to fire only ceramic tile bodies. The second firing was necessary to fire only the glaze, and it was generally carried out at lower temperature as a function of the type of glaze (≅700 °C).

Tables 1 and 2 summarize the energy consumption according to the phase of the production process and the type of product, respectively.

Table 1. Specific thermal and electrical consumption according to the process phase [40].

Process Phase	Specific Thermal Consumption [GJ/t]	Specific Electrical Consumption [GJ/t]
Wet grinding	-	0.05–0.35
Spray drying	1.1–2.2	0.01–0.07
Pressing	-	0.05–0.15
Drying	0.3–0.8	0.01–0.04
Firing	1.9–4.8	0.02–0.15

Table 2. Average total specific consumption according to the type of product [40].

Type of Product	Average Total Specific Consumption [GJ/t]
Single-firing tile	5.78–6.37
Double-firing tile	4.67

2.1. The Spray Dryer Technology

The spray dryer apparatus, as shown in Figure 3, essentially consists of a cylindrical chamber, where a heat exchange is promoted between the slip and a hot air flow with a temperature of about 500 °C. The slip is finely sprayed upward via nozzles in the form of droplets, while the hot air is directed tangentially downward to impose a spiral motion. Thanks to the heat exchange process, an immediate evaporation and consequent hardening of the external side of the droplet take place, and simultaneously, the water vapor in the interior leaves the droplet through its rear, which collapses inward, forming the characteristic hollow sphere appearance [38,39].

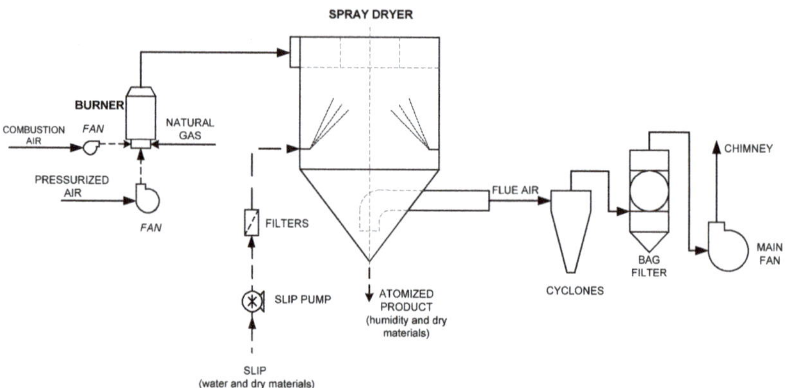

Figure 3. Schematic layout of the spray dryer unit.

The application of spray dryers in the ceramic tile industry has optimized the shaping process, assuring a high degree of compactness in the formed product. Spray-dried powders have a tailored particle size distribution (mainly concentrated in the 125–500 μm range) and a moisture content of ≅5–6%. The spray-dried powder optimally fills the press molds by the spherical shape of the granules, which improves its floatability and compaction. Furthermore, spray drying promotes the agglomeration of fine powders (<125 μm), avoiding de-airing problems during pressing [39].

The hot air flow in the spray dryer is usually generated by the combustion of natural gas and ambient air into a burner section (see Figure 3). At this stage, CHP units are usually installed, exploiting the gas turbine and internal combustion engine technologies.

As visible in Figure 3, the flue air stream, before being discharged into the atmosphere, is forced to pass through a gas cleaning section typically composed of cyclones and bag filters for dust removal.

In a large part of Italian installations, energy efficiency solutions are adopted to supply part of the heat required by the spray dryer. The study in [12] reports that over a sample of 71 spray dryers, 18% of them made use of heat recovery from kilns, 44% used CHP units, 4% used both systems, and 34% did not use any solution.

2.2. Cogeneration System Supporting the Dryer Process

Gas turbines (GTs) and internal combustion engines (ICEs) are the two CHP technologies currently installed in the ceramic tiles industry [12]. In both types of prime movers, the electrical energy generated by the CHP unit is used to fully or partially satisfy the electricity demand of the ceramic tile production process. The heat rejected by the prime mover (PM) is used to support the spray dryer process with the direct use of exhaust gases.

Simplified layouts of GTs and ICEs thermally integrated with a spray-dryer are presented in Figures 4 and 5, respectively.

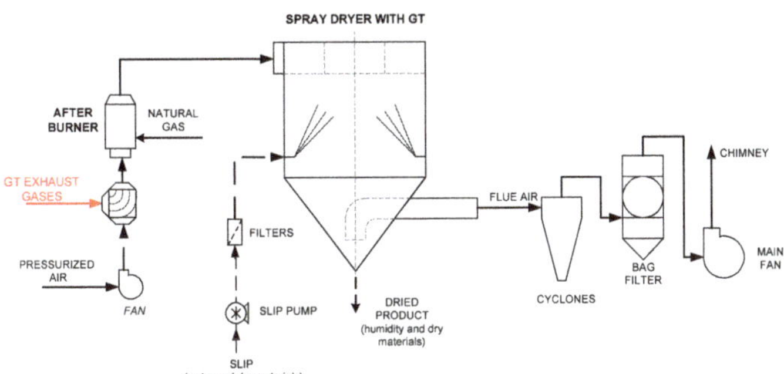

Figure 4. Schematic layout of the thermal integration between the combined heat and power (CHP) gas turbine (GT) unit and spray dryer.

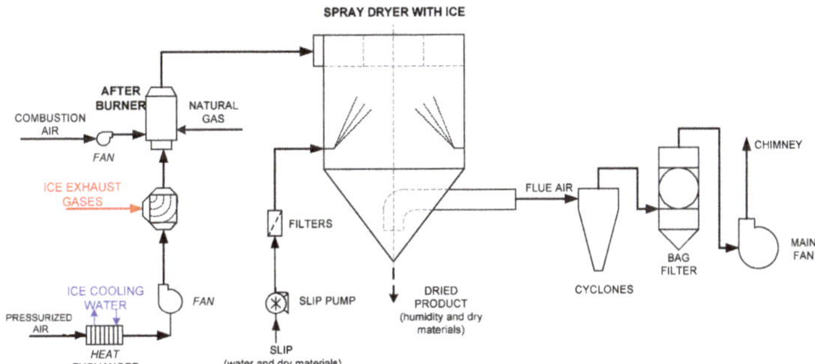

Figure 5. Schematic layout of the thermal integration between the CHP internal combustion engine (ICE) unit and spray dryer.

In both configurations, the CHP units' exhaust gases are heated up until the optimal process operating temperature (i.e., typically in the range 500–600 °C), thanks to an after-burner. The main difference between the arrangements lies in the amount of fresh air that is added into the process. Exploiting exhaust coming from a GT unit allows for the use of a small fraction of fresh air coming from ambient and pressurized by means of a fan ("pressurized air" in Figure 4). The streams are mixed, as in Figure 4, before entering the after-burner. If the mass flow of the exhaust is sufficient to support the drying process, the air mass flow could eventually be unnecessary. Conversely, the direct use of ICEs exhausted gases require a more significant amount of fresh air. Thus, as visible in Figure 5, both streams named "combustion air" and "pressurized air" are used in this configuration. The pressurized air stream is mixed with the ICE exhausted one and post combustion takes place, also introducing the combustion air inside the after-burner component.

This difference between the CHP spray dryer setup was mainly due to two reasons: (i) the different mass flow of GT and ICE exhaust gases discharged for a given electrical size, and (ii) the different concentration of oxygen in the exhaust. For the same electric size of the CHP unit, ICEs are characterized by a lower amount of thermal power discharged with the exhaust, both in terms of mass flow rate and temperature, due to the higher conversion efficiency compared to GTs, and to the fraction of low-temperature heat discharged to the engine coolant. In addition, GT exhausted gases are typically characterized by an oxygen concentration within 18–19% vol. dry, while the ICE typical oxygen concentration is lower, in between 14–15% vol. dry.

The last difference lies, as indicated in Figure 5, in the possibility of preheating the pressurized air stream exploiting the low-grade heat coming from the ICE water cooling circuit. As visible in Figure 5, a heat exchanger is placed upstream of the fan component to increase the temperature of the pressurized air stream exploiting the engine's cooling water heat. Typical air temperature increase is in the range 40–50 °C.

3. Methodology and Performance Indexes

In the ceramic sector, 197 energy audits carried out from 143 companies were collected by ENEA in December 2019 [4]. A total of 103 EAs from 68 companies referred to the tile ceramic process and have been analyzed to check the CHP presence in the industrial process. The total number of production sites including a CHP system was found equal to be 28.

Starting from the analysis EA collected by ENEA, a detailed assessment of the CHP units' specific characteristics was realized by integrating those data with information directly requested from the ceramic companies. For this purpose, a questionnaire was sent to all the ceramic tile industries equipped with CHP units and located in the tile district, containing the following specific macro-requests:

-   type and model of the CHP unit installed;
-   annual consumption of natural gas feeding the CHP unit and annual generated electricity;
-   annual operating hours of the CHP unit;
-   annual consumption of natural gas feeding the after-burner section;
-   annual amount of slip input to the dryer section and dried products generated;
-   annual operating hours of the dryer unit; and
-   annual electricity consumption of the ceramic tile production process.

In response to that, this study analyzed the annual operating data collected from ten CHP installations. The energetic analysis of the dryer process and of the CHP system was accomplished by introducing the following ad hoc-defined performance indexes:

CHP natural gas primary energy input, $Q_{in,CHP}$, calculated as the product of annual natural gas volume flow input to the CHP unit and lower heating value, assumed equal to 8250 kcal/Sm^3.

CHP average electric efficiency, $\overline{\eta_e}$, defined as the ratio between annual generated electrical energy, E_{CHP}, and primary energy input, $Q_{in,CHP}$.

Ratio of produced to consumed electrical energy, δ. This index helps to assess the design and the main target of the CHP unit. A value of δ equal to 1 means that all the energy generated by the CHP unit is used to satisfy the electricity consumption required by the production process. Values higher or lower than 1 mean, respectively, that a surplus or a deficit of electricity occurs compared to the production process needs.

Thermal energy discharged by the CHP unit, $Q_{gas,CHP}$. This variable accounts for the thermal energy discharged with PM exhaust gases, available to the thermal user. It has been evaluated, indirectly, based on following equations, according to the definition reported in [41]:

For the gas turbine:

$$Q_{gas,CHP} = Q_{in,CHP} \cdot \eta_1 - \frac{E_{CHP}}{\eta_2 \cdot \eta_3 \cdot \eta_4} \tag{1}$$

For the internal combustion engine:

$$Q_{gas,CHP} = Q_{in,CHP} \cdot \eta_1 \cdot k_5 - \frac{E_{CHP}}{\eta_2 \cdot \eta_3 \cdot \eta_4} \tag{2}$$

where $\eta_1, \eta_2, \eta_3,$ and η_4 in Equations (1) and (2) represent, respectively, the combustion chamber, auxiliaries, electric conversion, and gear box efficiencies. Conversion efficiency values have been assumed equal to the ones reported in Table 3. Coefficient k_5, only included in Equation (2), accounts for the fraction of discharged thermal energy (i.e., primary energy not converted into electricity) available in the high temperature heat circuit (i.e., with exhaust gases). The value of coefficient k_5, dependent on the ICE model, was assumed according to the manufacturers' specifications. Assumed values were in the range 0.74–0.78.

Table 3. Assumed value of conversion efficiencies [41].

η_1 [-]	η_2 [-]	η_3 [-]	η_4 [-]
0.9900	0.9800	0.9625	0.9850

To quantify the real amount of heat input to the spray dryer with PM exhaust gases, the quantity $Q_{gas,CHP}$ was corrected with a factor representing the ratio between the annual operating hours of the spray dryer unit and annual operating hours of the CHP system.

Thermal energy input to the dryer with natural gas, $Q_{in,SP}$, calculated as the product of annual natural gas volume flow input to the after-burner and lower heating value equal to 8250 kcal/Sm^3.

Total heat consumption of the dryer, Q_{TOT}. This variable accounts for the total heat input to the spray dryer component obtained by adding the two main thermal contributions $Q_{in,SP}$ and $Q_{gas,CHP}$. It must be pointed out that the heat requested to heat up and vaporize the water content in the slip represents the main contribution (typical values are in the range 70–80%) to the total process need. Minor contributions are represented by the heat necessary to heat up the combustion and/or the pressurized air streams, the heat absorbed by the solid matter, and the heat dissipated through the dryer walls as the process is not adiabatic.

Fraction of total heat consumption of the spray dryer covered with CHP discharged heat, Λ.

$$\Lambda = \frac{Q_{TOT}}{Q_{gas,CHP}} \tag{3}$$

Spray dryer specific consumption, C_s. This parameter is calculated as the ratio between Q_{TOT} and annual amount of evaporated water, $\dot{m}_{H2O}$, according to Equation (4):

$$C_s = \frac{Q_{TOT}}{\dot{m}_{H2O}} \tag{4}$$

The parameter C_s defines the amount of thermal energy requested by the process for unit mass of evaporated water, and it is normally expressed in kcal/kg H_2O. Alternatively, the specific consumption can also be expressed with reference to kg of dried product produced.

4. Energy Analysis Results and Discussion

This section is organized as follows. The first sub-section shows the results related to the CHP units, whereas the second sub-section describes the results related to the spray dryer as the thermal user.

4.1. Combined Heat and Power (CHP) Energetic Results

The preferred choice in terms of CHP prime mover is the internal combustion engine, with seven out of ten installations. Figure 6 shows the design electric power of the CHP prime movers plotted against the average annual mass flow rate of evaporated water. Blue squares in the figure represent the ICE units while the red circles represent GT units. The electric size of the installed CHP units was between 3.4 MW and 4.9 MW with an average value equal to 4 MW.

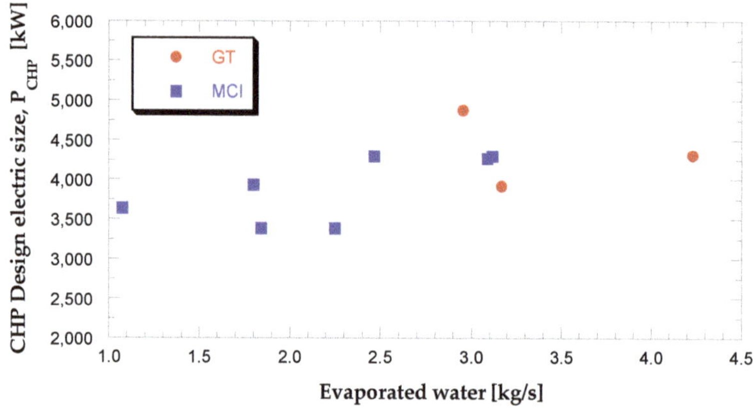

Figure 6. Design electric power size of the CHP units installed versus evaporated water average mass flow rate.

Gas turbines seem to be the preferred choice only when the size of the spray dryer technology is high (i.e., with average water mass flow rate higher than 2.50 kg/s), while ICEs are preferred with low evaporative capacity.

Figure 7 shows the electric efficiency values of installed CHP units plotted versus design electric size. ICE units are characterized by higher efficiency values, in the range 42–44%, compared to GTs whose values hardly exceed 30%. This outcome is also confirmed by the data reported in Figure 8, where the primary energy consumption of CHP units is plotted versus annual operating hours. Annual primary energy consumptions are in the range between 3210^6 and 10210^6 kWh/y, significantly higher for gas turbines compared to combustion engines for similar values of operating hours. Five CHP units reached operating hour values above 7000 h/y, three were between 6000 and 7000 h/y, while the remaining two were in operation for about half the time during the entire year (i.e., operating hours between 4200–4800 h/y).

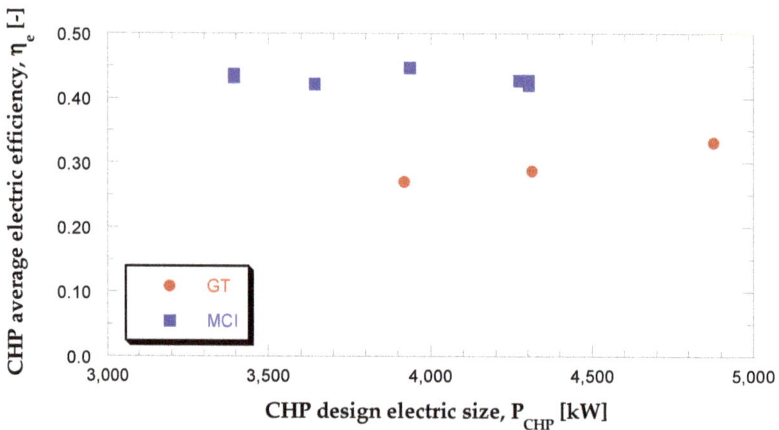

Figure 7. CHP average electric efficiency versus design electric size.

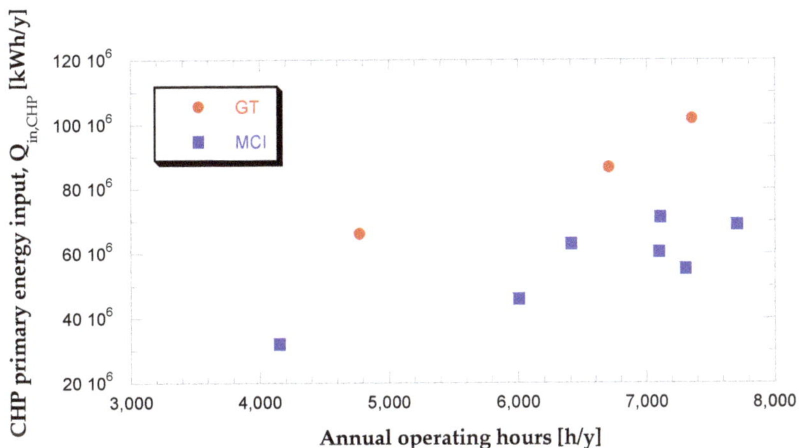

Figure 8. CHP natural gas primary energy consumption versus annual operating hours.

In Figure 9, the generated electrical energy is compared to the electricity demand of the facility. The black continuous line represents a value of δ equal to one (i.e., the case of self-consumption of the whole CHP generated electrical energy). As observed, the value of δ was higher than one in all of the analyzed cases. In particular, in six installations, the surplus of electricity was marginal (i.e., δ between 1.03 and 1.08), suggesting that the design and the load regulation strategy of the CHP units were targeted to meet the request of the facility. In the remaining cases, a modest surplus of electricity was generated (δ between 1.10 and 1.25). Only one case showed a significant amount of electricity generated compared to self-consumption (δ equal to 1.39).

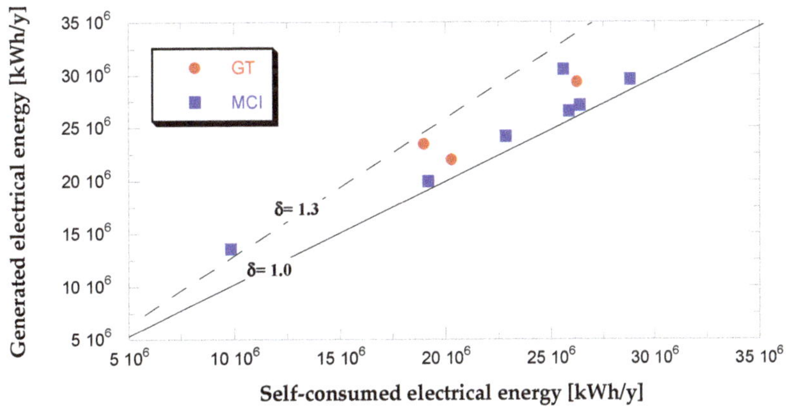

Figure 9. CHP generated electrical energy versus the electricity demand of the facility.

4.2. Energetic Results of CHP-Spray Dryer Integrated System

The total heat consumptions of the spray dryer units are plotted in Figure 10 as the function of annual evaporation capacity. As expected, an increase in thermal energy required for the drying process was observed as the evaporative capacity increased. The obtained trend was almost linear with values of Q_{TOT} in the range 1510^6–7510^6 kWh/y.

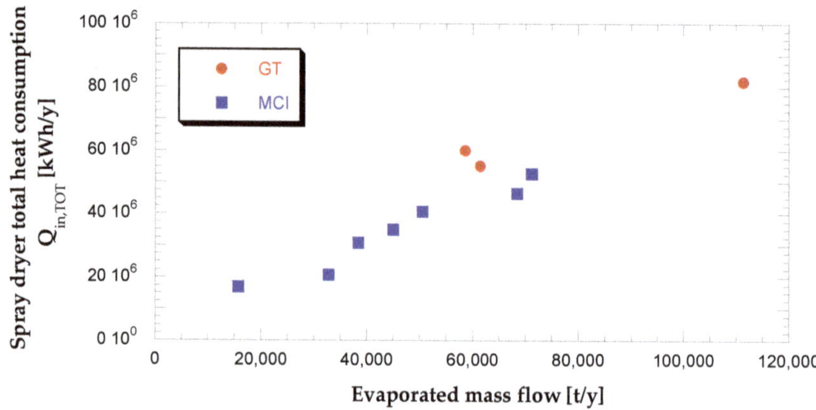

Figure 10. Total heat consumption of the spray dryer plotted against evaporated mass flow.

Contributions of CHP thermal energy, $Q_{gas,CHP}$, and natural gas energy input, $Q_{in,SP}$, to the dryer's heat consumption are shown in Figures 11 and 12, respectively. GTs showed a higher value of $Q_{gas,CHP}$ (between 4010^6–6010^6 kWh/y) compared to the ICEs, thus confirming that their limited electrical efficiency values (see results in Figure 8) had a positive effect on the amount of discharged heat. Percentage values of Λ for GT units, as shown in Figure 13, were between 67% and 77%. In contrast, in the case of ICEs, the contribution of thermal energy discharged with exhaust was limited ($Q_{gas,CHP}$, in between 810^6 and 1410^6 kWh/y), resulting in Λ values between 30 and 45%. It must be highlighted that the value of Λ seems to be dependent only on the technology used as the CHP prime mover, not being influenced by the size of the spray dryers (i.e., amount of evaporated water).

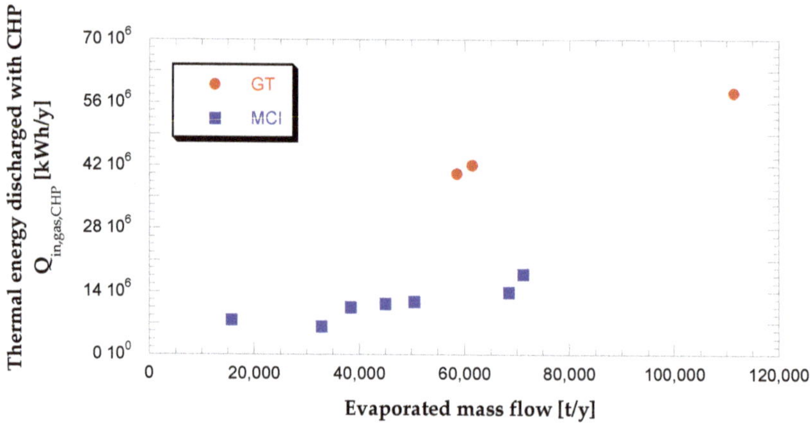

Figure 11. Thermal energy discharged with CHP and input to spray dryer versus the evaporated mass flow.

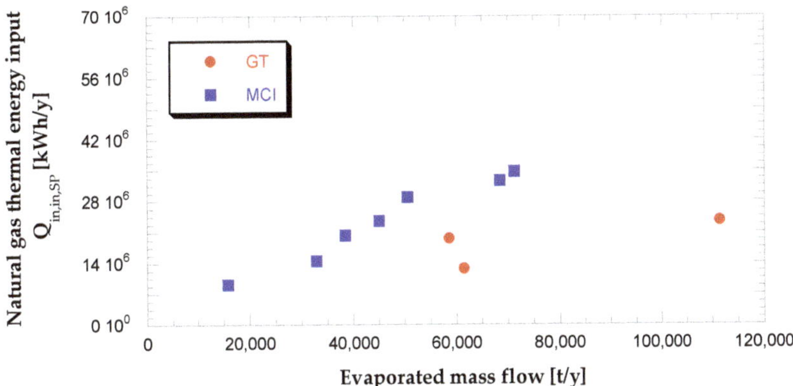

Figure 12. Thermal energy input to spray dryer with natural gas versus evaporated mass flow.

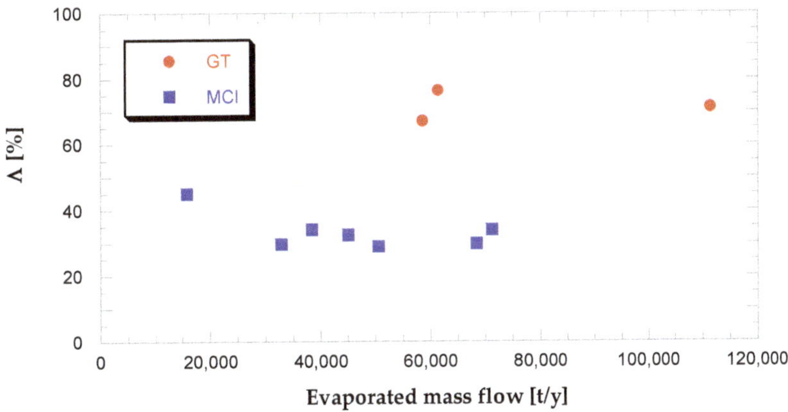

Figure 13. Percentage of total heat consumption of the spray dryer covered with CHP discharged heat versus evaporated mass flow.

Based on the presented results, it can be asserted that for a given prime mover's size, the choice of ICE guarantees the maximum generation of electricity, while the GT allows for the minimization of the consumption of natural gas required to support the spray dryer thermal process. It must be pointed out that other factors such as the engines' flexibility, maintenance costs, specific emission values, etc., can of course affect the choice of the prime mover technology to be installed.

Finally, the spray dryers' specific consumption is presented in Figure 14, referred to as the mass of evaporated water (Figure 14a,b) and to the mass of dried product.

The obtained values were in the range of 600–900 kcal/kg_{H2O} and 250–420 kcal/kg_{dried}, in line with values reported in [37]. Average values were equal to 710 kcal/kg_{H2O} and 305 kcal/kg_{dried}.

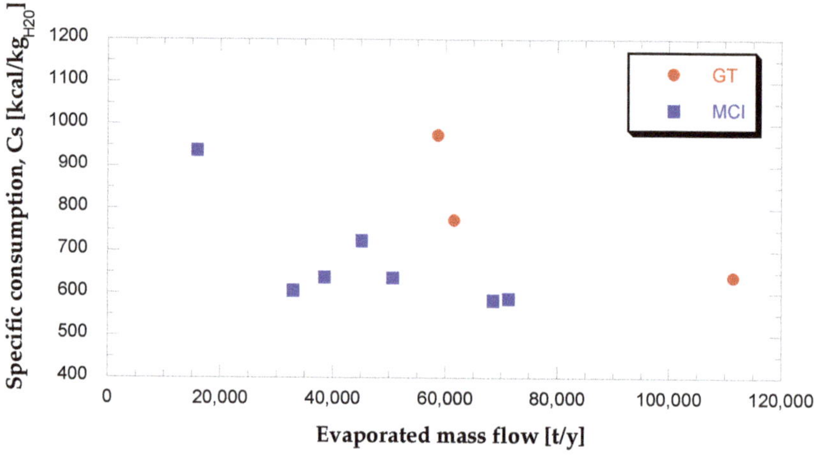

(a)

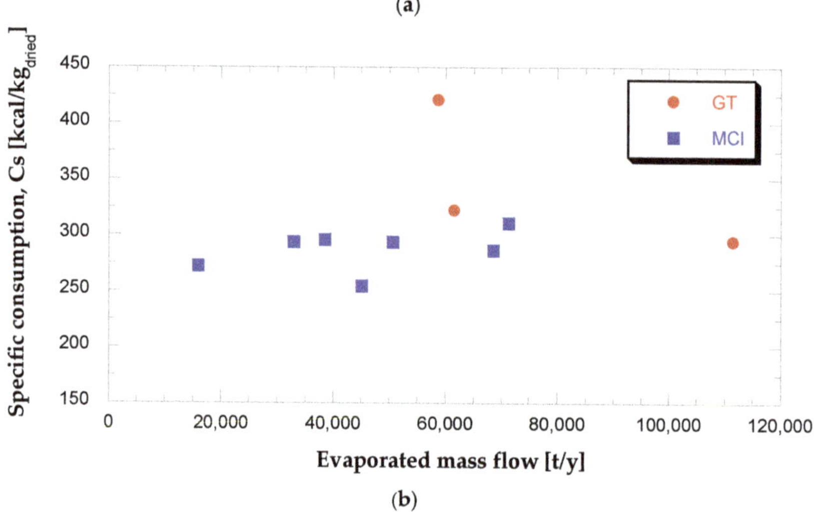

(b)

Figure 14. Spray dryer thermal specific consumption based on evaporated mass flow (**a**) and dried product (**b**).

5. Conclusions

In this study, the preliminary results of a two-year research project named "Energy efficiency of industrial products and processes", aimed at evaluating the benefits of cogeneration (CHP) applied to the ceramic tile industry, are presented.

Then findings demonstrate how the use of CHP technology can contribute to pursue the energy savings targets intended within the energy efficiency policy measures. Installed CHP units are used to generate onsite electricity, while thermal energy discharged with exhaust gases is directly used to supply the spray dryer, thus reducing the consumption of natural gas.

Obtained results of the study showed that:

- The electric size of the installed CHP units was between 3.4 MW and 4.9 MW, with an average value equal to 4 MW.
- CHP installed prime movers are internal combustion engine and gas turbines.

- Internal combustion engines are often the preferred choice, due to higher conversion efficiency values in electricity production (found in the range 42-44%).
- In contrast, GTs seem to be the preferred choice only when the size of the spray dryer unit is high (i.e., with evaporated mass flow rate higher than 2.5 kg/s) and the target is to minimize the consumption of natural gas input to the dryer.
- The total specific consumption of the spray dryer process was quantified in the range 600–950 kcal/kg_{H2O} or 250–420 kcal/kg of dried product, in line with values reported in the specific literature.
- The percentage of specific consumption covered with CHP thermal energy is strictly dependent on the type of prime mover installed: lower values, in the range 30–45%, are characteristic of combustion engines, whereas the use of gas turbines can contribute up to 77% of the process's total consumption.

The results aim to provide an overview of the current CHP installations in the Italian ceramic tile sector and could also be used as guidelines in the selection and design of CHP units to industrial developers.

Future steps of the research activity will be focused on thermodynamic modeling of the integrated system for a detailed representation of the thermal fluxes involved. Moreover, a methodology will be defined to assess the reduction of the environmental impact related to the use of a cogeneration unit in the ceramic tile industry.

Author Contributions: Conceptualization, L.B.; Formal analysis, L.B., B.F., and B.M.; Methodology, L.B.; Resources A.C.; Data curation and validation B.F., B.M., F.M., and M.S.; Supervision, M.C.B. and F.M.; Writing—original draft, L.B., B.F., B.M., F.M., M.S., and C.T.; Writing—review & editing, S.O.; Project administration M.C.B., F.M., M.S., and C.T. All authors have read and agreed to the published version of manuscript.

Funding: The research activity was funded by the Electrical System Research (PTR 2019-2021), implemented under Program Agreements between the Italian Ministry for Economic Development and ENEA, CNR, and RSE S.p.A.

Conflicts of Interest: The authors declare no conflict of interest.

References

1. European Roadmap 2050. Available online: https://www.roadmap2050.eu/ (accessed on 15 February 2021).
2. Gerbaulet, C.; von Hirschhausen, C.; Kemfert, C.; Lorenz, C.; Oei, P.-Y. European electricity sector decarbonization under different levels of foresight. *Renew. Energy* **2019**, *141*, 973–987. [CrossRef]
3. Directive 2012/27/EU of the European Parliament and of the Council of 25 October 2012 on Energy Efficiency. Available online: https://eur-lex.europa.eu/legal-content/EN/TXT/?uri=celex%3A32012L0027 (accessed on 5 February 2021).
4. ENEA. Rapporto Annuale Efficienza Energetica 2020. Available online: https://www.efficienzaenergetica.enea.it/component/jdownloads/?task=download.send&id=453&catid=40%20&Itemid=101 (accessed on 20 October 2020).
5. International Energy Agency (IEA). Heating without Global Warming—Market Developments and Policy Considerations for Renewable Heat. 2014. Available online: https://www.iea.org/reports/heating-without-global-warming (accessed on 5 February 2021).
6. International Energy Agency (IEA). World Energy Outlook 2020. Available online: https://www.iea.org/reports/world-energy-outlook-2020 (accessed on 5 February 2021).
7. International Energy Agency (IEA). Clean and Efficient Heat for Industry. Available online: https://www.iea.org/commentaries/clean-and-efficient-heat-for-industry (accessed on 5 February 2021).
8. Bilancio Energetico Nazionale. Available online: https://dgsaie.mise.gov.it/ben.php (accessed on 25 February 2021).
9. TERNA S.P.A. Dati Statistici sull'Energia Elettrica in Italia, Annuario Statistico 2018. Available online: https://download.terna.it/terna/Annuario%20Statistico%202018_8d7595e944c2546.pdf (accessed on 5 February 2021).
10. Energy & Strategy Group. *Energy Efficiency Report*, 9th ed.; Available online: https://www.energystrategy.it/area-riservata/efficiency-report.html (accessed on 25 February 2021).
11. Ministry of Economic Development. Relazione Annuale Sulla Cogenerazione in Italia. April 2019. Available online: https://www.mise.gov.it/images/stories/documenti/Relazione-annuale-sulla-CAR-2019.pdf (accessed on 25 February 2021).
12. Confindustria Ceramica. Indagine Sugli Interventi di Efficientamento Termico Nel Settore delle Piastrelle di Ceramica. 2016. Available online: file:///C:/Users/utente/Desktop/download/5759ISN2018_PIASTREL.pdf (accessed on 10 December 2020).

13. Confindustria Ceramica. Indagini Statistiche sull'industria Italiana. 2020. Available online: http://www.confindustriaceramica.it/site/en/home/bookstore/studi-di-settore/indagine-statistica-sullindustria-italiana----piastrelle-di-ceramica.html (accessed on 10 December 2020).

14. Bindlish, R. Scheduling, optimization and control of power for industrial cogeneration plants. *Comput. Chem. Eng.* **2018**, *114*, 221–224. [CrossRef]

15. Çakir, U.; Çomakli, K.; Yüksel, F. The role of cogeneration systems in sustainability of energy. *Energy Convers. Manag.* **2012**, *63*, 196–202. [CrossRef]

16. Armanasco, F.; Colombo, L.P.M.; Lucchini, A.; Rossetti, A. Techno-economic evaluation of commercial cogeneration plants for small and medium size companies in the Italian industrial and service sector. *Appl. Therm. Eng.* **2012**, *48*, 402–413. [CrossRef]

17. Gambini, M.; Vellini, M. High Efficiency Cogeneration: Performance Assessment of Industrial Cogeneration Power Plants. *Energy Procedia* **2014**, *45*, 1255–1264. [CrossRef]

18. Tsai, W.; Hsien, K. An analysis of cogeneration system utilized as sustainable energy in the industrial sector in Taiwan. *Renew. Sustain. Energy Rev.* **2007**, *11*, 2104–2120. [CrossRef]

19. Soares, J.B.; Szklo, A.S.; Tolmasquim, M.T. Incentive policies for natural gas-fired cogeneration in Brazil's industrial sector—Case studies: Chemical plant and pulp mill. *Energy Policy* **2001**, *29*, 205–215. [CrossRef]

20. Bianco, V.; De Rosa, M.; Scarpa, F.; Tagliafico, L.A. Implementation of a cogeneration plant for a food processing facility. A case study. *Appl. Therm. Eng.* **2016**, *102*, 500–512. [CrossRef]

21. Khurana, S.; Banerjee, R.; Gaitonde, U. Energy balance and cogeneration for a cement plant. *Appl. Therm. Eng.* **2002**, *22*, 485–494. [CrossRef]

22. Panno, D.; Messineo, A.; Dispenza, A. Cogeneration plant in a pasta factory: Energy saving and environmental benefit. *Energy* **2007**, *32*, 746–754. [CrossRef]

23. Freschi, F.; Giaccone, L.; Lazzeroni, P.; Repetto, M. Economic and environmental analysis of a trigeneration system for food-industry: A case study. *Appl. Energy* **2013**, *107*, 157–172. [CrossRef]

24. Ruiz Celma, A.; Cuadros Blazquez, F.; Lopez-Rodriguez, F. Feasibility analysis of CHP in an olive processing industry. *J. Clean. Prod.* **2013**, *42*, 52–57. [CrossRef]

25. Tang, O.; Mohanty, B. Industrial energy efficiency improvement through cogeneration: A case study of the textile industry in Thailand. *Energy* **1996**, *21*, 1169–1178. [CrossRef]

26. Caglayana, H.; Caliskanb, H. Energy, exergy and sustainability assessments of a cogeneration system for ceramic industry. *Appl. Therm. Eng.* **2018**, *136*, 504–515. [CrossRef]

27. Caglayana, H.; Caliskanb, H. Thermodynamic based economic and environmental analysis of an industrial cogeneration system. *Appl. Therm. Eng.* **2019**, *158*. [CrossRef]

28. Caglayana, H.; Caliskanb, H. Advanced exergy analyses and optimization of a cogeneration system for ceramic industry by considering endogenous, exogenous, avoidable and unavoidable exergies under different environmental conditions. *Renew. Sustain. Energy Rev.* **2021**, *140*, 110730. [CrossRef]

29. Yoru, Y.; Karakoc, T.H.; Hepbasli, A. Dynamic energy and exergy analyses of an industrial cogeneration system. *Int. J. Energy Res.* **2009**, *34*, 345–356. [CrossRef]

30. Delpech, B.; Milani, M.; Montorsi, L.; Boscardin, D.; Chauhan, A.; Almahmoud, S.; Axcell, B.; Jouhara, H. Energy efficiency enhancement and waste heat recovery in industrial processes by means of the heat pipe technology: Case of the ceramic industry. *Energy* **2018**, *158*, 656–665. [CrossRef]

31. Jouhara, H.; Bertrand, D.; Axcell, B.; Montorsi, L.; Venturelli, M.; Almahmoud, S.; Milani, M.; Ahmad, L.; Chauhan, A. Investigation on a full-scale heat pipe heat exchanger in the ceramics industry for waste heat recovery. *Energy* **2021**, *223*, 120037. [CrossRef]

32. Mezquita, A.; Boix, J.; Monfort, E.; Mallol, G. Energy saving in ceramic tile kilns: Cooling gas heat recovery. *Appl. Therm. Eng.* **2014**, *65*, 102–110. [CrossRef]

33. Agrafiotis, C.; Tsoutsos, T. Energy saving technologies in the European ceramic sector: A systematic review. *Appl. Therm. Eng.* **2001**, *21*, 1231–1249. [CrossRef]

34. Ros-Dosdà, T.; Fullana-i-Palmer, P.; Mezquita, A.; Masoni, P.; Monfort, E. How can the European ceramic tile industry meet the EU's low-carbon targets? A life cycle perspective. *J. Clean. Prod.* **2018**, *199*, 554–564. [CrossRef]

35. Almeida, M.I.; Dias, A.C.; Demertzi, M.; Arroja, L. Environmental profile of ceramic tiles and their potential for improvement. *J. Clean. Prod.* **2016**, *131*, 583–593. [CrossRef]

36. Boschi, G.; Masi, G.; Bonvicini, G.; Bignozzi, M.C. Sustainability in Italian Ceramic Tile Production: Evaluation of the Environmental Impact. *Appl. Sci.* **2020**, *10*, 9063. [CrossRef]

37. Nassetti, G.; Ferrari, A.; Fregni, A.; Maestri, G. *Piastrelle Ceramiche e Energia: Banca Dati dei Consumi Energetici nell'Industria delle Piastrelle di Ceramica*; Assopiastrelle: Sassuolo, Italy, 1998.

38. Maroncelli, M.; Timellini, G.; Evangelisti, R. *I Consumi Energetici nella Produzione delle Piastrelle Ceramiche*; Centro Ceramico: Bologna, Italy, 1985.

39. SACMI. *Applied Ceramic Technology*; Editrice la Mandragora: Imola, Italy, 2002; Volume 2.

40. Ministro Dell'ambiente e Della Tutela Del Territorio e Del Mare. Decreto Interministeriale 29 Gennaio 2007: "Emanazione di Linee Guida per l'Individuazione e Applicazione delle Migliori Tecniche Disponibili in Materia di Vetro, Fritte Vetrose e Prodotti Ceramici, per le Attività Elencate nell'Allegato I del Decreto Legislativo 18 Febbraio 2005, Numero 59" Gazzetta Ufficiale 31 maggio 2007, n. 125, S.O.
41. Gestore dei Servizi Energetici (GSE). Guida alla Cogenerazione ad Alto Rendimento CAR, Rev. 2019. Available online: https://gse.it/documenti_site/Documenti%20GSE/Servizi%20per%20te/COGENERAZIONE%20AD%20ALTO%20RENDIMENTO/Guide/Aggiornamento%20Guida%20CAR%20-%20revisione%202019.pdf (accessed on 1 March 2021).

 *sustainability*

Article

Technological Energy Efficiency Improvements in Cement Industries

Alessandra Cantini [1], **Leonardo Leoni** [1], **Filippo De Carlo** [1,*], **Marcello Salvio** [2], **Chiara Martini** [2] and **Fabrizio Martini** [2]

1   Department of Industrial Engineering (DIEF), University of Florence, Viale Morgagni, 40, 50134 Florence, Italy; alessandra.cantini@unifi.it (A.C.); leonardo.leoni@unifi.it (L.L.)
2   DUEE-SPS-ESE Laboratory, Italian National Agency for New Technologies, Energy and Sustainable Economic Development (ENEA), Lungotevere Thaon di Revel, 76, 00196 Rome, Italy; marcello.salvio@enea.it (M.S.); chiara.martini@enea.it (C.M.); fabrizio.martini@enea.it (F.M.)
*   Correspondence: filippo.decarlo@unifi.it; Tel.: +39-05-5275-8677

Abstract: The cement industry is highly energy-intensive, consuming approximately 7% of global industrial energy consumption each year. Improving production technology is a good strategy to reduce the energy needs of a cement plant. The market offers a wide variety of alternative solutions; besides, the literature already provides reviews of opportunities to improve energy efficiency in a cement plant. However, the technology is constantly developing, so the available alternatives may change within a few years. To keep the knowledge updated, investigating the current attractiveness of each solution is pivotal to analyze real companies. This article aims at describing the recent application in the Italian cement industry and the future perspectives of technologies. A sample of plant was investigated through the analysis of mandatory energy audit considering the type of interventions they have recently implemented, or they intend to implement. The outcome is a descriptive analysis, useful for companies willing to improve their sustainability. Results prove that solutions to reduce the energy consumption of auxiliary systems such as compressors, engines, and pumps are currently the most attractive opportunities. Moreover, the results prove that consulting sector experts enables the collection of updated ideas for improving technologies, thus giving valuable inputs to the scientific research.

Keywords: cement manufacturing plant; energy savings; technology solutions; Italian companies; descriptive analysis

Citation: Cantini, A.; Leoni, L.; De Carlo, F.; Salvio, M.; Martini, C.; Martini, F. Technological Energy Efficiency Improvements in Cement Industries. *Sustainability* **2021**, *13*, 3810. https://doi.org/10.3390/su13073810

Academic Editor: Adam Smoliński

Received: 23 February 2021
Accepted: 26 March 2021
Published: 30 March 2021

Publisher's Note: MDPI stays neutral with regard to jurisdictional claims in published maps and institutional affiliations.

1. Introduction

The production process in cement manufacturing plants is typically energy-intensive and requires large amounts of resources [1]. A typical well-equipped plant consumes about 4 GJ of energy to produce one ton of cement, while the cement production in the world is about 3.6 billion tons per year [2]. It was estimated that the cement manufacturing process consumes around 7% of industrial energy consumption, which, in turn, accounts for 30–40% of the global energy consumption [3].

Given the significant impact that the manufacturing industry has on global sustainability and considering the increasing economic pressure introduced by a competitive market and the reduction of available energy resources, optimizing the energy efficiency of production systems has become a primary concern [1]. For this purpose, to reduce energy consumption in the cement sector, it is possible to act both on a technological and a managerial level [4]. Focusing on the technological aspects, one of the strategies to be embraced is to improve production plants by modifying or replacing inefficient equipment with better-performing and less energy-intensive ones [5,6].

As stated by Su et al. [7], the cement manufacturing process can be divided into three major stages: raw material processing, clinker production, and finish grinding processing (finished cement production). A schematic representation of the cement production process is illustrated in Figure 1. The raw material processing reduces the size of limestone and clay extracted from the quarries, thus obtaining a homogeneous mixture with an appropriate chemical composition. At first, one or more subsequent crushers break down the raw material, reducing the rocks' size from 120 cm to 1.2–8 cm [8]. The crushed rocks are then pre-blended through dedicated apparatuses called stackers and reclaimers. Next, the ingredients are dosed and fed to the mills, which perform the grinding process. During this phase, rocks are ground to fine particles, moved through mechanical conveyors, or fluidized channels fed by blowers, and transported to the homogenizing silos, where the blending process takes place. As the last step of raw material processing, the blending process allows obtaining a uniform chemical composition. Before entry into the kiln, the homogenized ingredients are sent to the preheater tower, where they flow through a series of cyclones. Here, fine particles are preheated exploiting the kiln's exhaust gases, reducing the energy required to carry out the subsequent heating process. The heating process is performed by the kiln, which could be arranged vertically or horizontally. In the kiln, the material temperature is increased over 1000 °C (sometimes up to 1400 °C), resulting in the formation of calcium silicate crystals-cement clinker [8]. At the kiln exit, the clinker is directed to a cooler and lowered in temperature. The cooling process is required to stop the chemical reactions at the right moment, allowing to obtain a proper quality of the products. Moreover, the cooling process allows to recover some heat from the hot clinker. Finally, the cooled clinker is stored inside silos and then fed to specific mills for the finish grinding process. During the final milling phase, some substances such as fly ash, limestone, slag, gypsum, and pozzolana are added to the clinker, depending on the requirements of the final product. In addition to the aforementioned processes, during the clinker production stage, two other complementary processes are performed: fuel preparation and exhaust gas treatment. Indeed, the fuel (i.e., pet-coke, carbon coke, Refuse-Derived Fuel—RDF, or others) must be milled before entering the kiln, while exhaust gases must be treated to remove dust and reduce emissions.

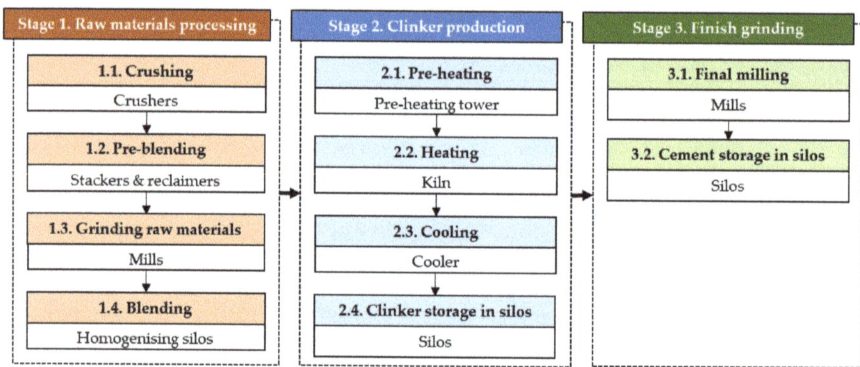

Figure 1. Cement production process. The colored boxes show the sequential operations, while the white boxes summarize the process machinery.

Based on the water content of the raw materials, the cement manufacturing process can be divided into four categories: dry, semi-dry, semi-wet, and wet [9]. Since cement production requires the complete evaporation of the water in raw ingredients, the higher the percentage of water, the more energy-intensive the process will be. On the other side, a higher water content facilitates the processes of homogenizing and blending. It is worth mentioning that several cement plants only deal with the final milling phase. In these plants, the starting material is the clinker, which has already undergone the heating process. Therefore, for such manufacturing sites, the above classification is not valid.

Generally, energy consumption in the cement industry is provided by electricity and fuels. Over 90% of fuels used are consumed in the production of clinker. Electric energy, on the other hand, is used for about 39% for the finishing process, for around 28% for both processing the raw materials and burning the clinker, and for less than 5% for other operations [7].

To pursue greater energy efficiency in the cement manufacturing plants, for each operation forming the production process, a wide variety of technological solutions has been developed. For instance, solutions to reduce energy problems during raw material crushing are described in [4,10]. Ref. [9] shows solutions for the slurry production process, Ref. [11] for pre-homogenization, and Ref. [12] for drying. Technological opportunities are also shown for grinding [13], homogenization [14], granulation [9], preheating [15], heating [16], cooling [17], and final grinding [18]. Other measures were outlined to improve the efficiency of dedusting [9], exhaust gas treatment, and heat recovery [4,19]. Finally, multiple solutions were developed for material transport systems [20] and auxiliary services [21].

Nevertheless, the multitude of more or less efficient technological alternatives simultaneously offered on the market hamper the choice of which improvement intervention is better to adopt. Therefore, in the scientific literature, several efforts have already been made to develop a review of the main technologies for overcoming energy problems and minimizing energy consumption [9,17,22,23]. Decision support tools have also been developed to help companies in selecting optimal energy efficiency measures [24]. In this context, a crucial role is played by the Best Available Techniques Reference Documents (BREFs), a series of documents developed by the European Union to review and describe industrial processes, their respective operating conditions, pollution factors, and strategies for improving industrial sustainability. In the case of the cement sector, the latest BREFs were published in 2013, in which, in addition to paying particular attention to pollution issues, the main strategies to reduce plant energy consumption were also detailed, considering both technological and technical (management and process control) solutions [4].

A synthetic classification that summarizes the main technological energy-saving alternatives, also dividing them according to the specific phase and asset on which they act, is provided in Appendix A (Tables A1–A17, black writings). It is worth mentioning that solutions related to auxiliary and heat recovery systems, being used in multiple process stages, were not associated with a specific manufacturing operation, but were considered as a separate category.

As technology is constantly developing [23], the available alternatives may change within a few years. Some new solutions may emerge, and others may fall into disuse due to their inferior performance. To this end, in addition to consulting the literature, a good way to update the state of the art of energy-saving technological opportunities and to check their degree of application is to consult a sample of companies working in the cement industry [19,25]. Due to the increasing competitiveness in the global industrial sector [26], companies are constantly looking for better technological solutions. Consequently, their consultation could be useful to keep up to date with energy efficiency measures currently attractive in real-life contexts. In this perspective, many studies were carried out on Chinese industries. For instance, the spread of technologies in Taiwanese plants is described by Huang et al. [25], while the current status of energy-efficiency opportunities in Shandong industries is shown by Hasanbeigi et al. [27]. On the contrary, to the best of the authors' knowledge, a similar study on Italian industries is missing.

As stated by Supino et al. [28], Italy and Germany are the most important cement manufacturers in Europe. In 2019, cement production in Italy was 19,240.645 tons [29]. The Italian cement sector is characterized by non-uniform operators, including multinational groups and other smaller and medium enterprises that operate both at national and local levels. In response to a period of crisis, the Italian cement industry radically changed its energy mix over the last two decades. The contribution of natural gas (-69%) and heavy fuel oil (-60%) was reduced, while the proportion of alternative fuels (such as RDF) was increased. Given the central role of the Italian cement industry in Europe, the diversification of its companies, and the energy progress occurring in such country [30], the results of the Italian analysis could also give important information on the European scenario.

To fill the gap caused by the absence of an Italian study, the contribution of the present paper is to show the Italian current situation and future perspectives of energy-saving technologies for cement plants, thus expanding the knowledge on energy efficiency measures. In this paper, a sample of Italian companies was analyzed consulting the results of mandatory energy audit and collecting information on which of the energy efficiency measures were recently implemented in each company and which will be applied in the next few years. Thus, in the present work, we aim to answer three research questions.

RQ1: *What technological energy-saving interventions were applied by Italian cement companies in the last four years?*

RQ2: *What measures are suggested by the same sample of companies to be applied in the next four years?*

RQ3: *What are the reasons why some solutions are preferred to others? What conclusions about the future development of Italian companies can be drawn by analyzing a sample of companies?*

The importance of the present study is to provide a realistic representation of the current Italian scenario. The results of the analysis are used both to extend Tables A1–A17 (Appendix A), indicating other possible measures in addition to those found in the literature, and, above all, to identify through statistical analysis the current degree of implementation of technological interventions. A coherent and systematic reflection on the diffusion of technologies in recent years provides a significant insight into their level of maturity and attractiveness for companies [25].

The methodology proposed in this paper is of significant importance because, by investigating a sample of real companies, the spread level and current situation of technologies in one of the most important cement manufacturers in Europe (Italy) is shown. The outcome of this document could be useful for companies willing to improve their sustainability by implementing new technological opportunities. Moreover, it could be useful to gain a better understanding of the Italian industry context. Finally, the consultation of sector experts is important because it allows collecting updated ideas for improving technologies, thus giving valuable inputs to the scientific research.

Focusing on technological opportunities, this article refers only to solutions concerning cement production equipment, excluding managerial energy-saving measures and measures related, for example, to the use of waste fuels. Moreover, the discussion excludes solutions concerning the lighting or heating of the industrial shed where the plant is located [31], solutions concerning quality control of finished products and packaging activities (which are downstream of the production process) [17], and solutions concerning the installation of sensors which, despite being technologies, are typically used together with software and computer systems to implement management solutions [6].

The remainder of the present paper is organized as follows: Section 2 outlines the approach followed to reach the goals; Section 3 offers the application of the proposed methodology and the description of the Italian scenario; and finally, Sections 4 and 5 provide a discussion on the results and some conclusions.

2. Materials and Methods

To investigate the actual state of cement production in Italy, an approach similar to the one presented by Hasanbeigi et al. [27] for analyzing Chinese provinces was adopted. The approach was focused on data collection to characterize the cement sector at a national level, thus obtaining an overview of the Italian scenario. This work was carried out in collaboration with the Italian cement trade association (Federbeton) and with the national agency for new technologies, energy, and sustainable economic development (ENEA). The aim of this approach was to analyze a sample of Italian companies, providing an insight into their current interest in energy-saving technologies and their future development directions.

The Energy Efficiency Directive 2012/27/EU (EED) is a solid cornerstone of Europe's energy legislation. It includes a balanced set of binding measures planned to help the EU reach its 20% energy efficiency target by 2020. The EED establishes a common framework of measures for the promotion of energy efficiency (EE) to ensure the achievement of the European targets and to pave the way for further EE improvements beyond 2020. The Italian Government transposed the EED in 2014 (by issuing the Legislative Decree n. 102/2014, recently updated by Legislative Decree n. 73/2020), extending the obligation also to a specific group of energy-intensive enterprises (mostly SMEs) and assigning ENEA (Italian National Agency for New Technologies, Energy, and the Sustainable Economic Development) the management of EED Article 8 obligation [32], where Article 8 is dedicated to Energy Audits, a necessary tool to assess the existing energy consumption and identify the whole range of opportunities to save energy.

In the EED, Energy Audit is defined as a systematic procedure aimed at obtaining adequate knowledge on the existing energy consumption profile of a building or group of buildings, an industrial or commercial operation or installation or private or public service, identifying, and quantifying cost-effective energy savings opportunities, and reporting the findings.

According to Art. 8 of Lgs. D. 102/14, two categories of companies have been targeted as obliged to carry out energy audits on their production sites, firstly by the 5th of December 2015, and then at least every four years: large enterprises and energy-intensive enterprises. In Italy, an organization qualifies as a large enterprise if it shows one of the following characteristics:

- Number of employees ≥250 and annual turnover >€ 50 million and annual budget >€ 43 million;
- Number of employees ≥250 and annual turnover >€ 50 million;
- Number of employees ≥250 and annual budget >€ 43 million.

Under the implementation of Article 8 in Italy, the size of the company must be calculated only on Italian sites both of the company itself and of its associated/related companies. Energy Intensive Enterprises are the ones with large energy consumptions applying for tax relief on part of the purchased energy. All the energy-intensive enterprises are registered on the list of «Cassa per i servizi energetici ed ambientali» (Governamental Agency related to electricity). Obliged Enterprises that will not carry out an energy audit observing Annex II of the EED within the above deadlines will be subject to administrative monetary penalties. The penalty does not exempt obliged enterprises from carrying out the audit, with its submission to ENEA within six months from the sanction imposition by the Ministry of Economic Development. According to Article 8 of Italian Legislative Decree 102/2014 implementing the Energy Efficiency Directive, as of 31 December 2019, ENEA received 11,172 energy audits of production sites relating to 6434 companies.

Over 53% of the audits were carried out on sites related to the manufacturing sector (8% Plastic, 9% Iron and Steel, 2% Paper, 3% Textile, 6% Food Industry) and over 14% in trade. Seventy percent of the audits collected by ENEA are equipped with specific monitoring of energy consumption.

In the cement sector, 65 Energy Audits carried out from 27 companies were collected by ENEA in December 2019 [33] and 61 energy Audits were characterized by the presence of a monitoring plan for energy consumption (11 companies are equipped with the ISO 50001 energy management system).

Each Energy Audit contains specific information on the geographical location of the plant, some general characteristics of the company, the type of production process, and the type of finished products. In addition, the energy audit is useful to collect additional information on:

- Which energy efficiency measures were implemented by the company in the last four years;
- Which were planned to be implemented in the next four years;
- Why they applied or they are suggesting such solutions.

These additional information (not referring to Tables A1–A17, Appendix A) were deliberately analyzed for two reasons. On the one hand, to confirm the validity and the attractiveness of the solutions already found in the literature, in the case they were proposed by companies. On the other hand, to extend and update the list of solutions if sector experts proposed alternative technologies to those already identified.

From 65 energy audits, a sample of 48 Italian cement plants was selected. The gathered information was registered in a database (Excel spreadsheet) and used to expand the list of technological solutions (see Tables A1–A17 in Appendix A, red writings). Then, the generated list was shared with the cement trade association (Federbeton) to obtain observations from industry experts and validate the output. Appendix A provides companies with a synthetic tool to improve their sustainability in each specific process phase or cement production machine.

Moreover, the collected data was analyzed, checking which interventions from the list were proposed and by how many companies. The following statistical parameters were computed for each alternative on the list:

- Number of companies that applied the i-th intervention in the last four years ($n_{a,i}$);
- Number of companies that suggested i as a future action ($n_{s,i}$);
- Frequency of application of i in the last four years ($f_{a,i}$);
- Frequency of suggestion of i for the next four years ($f_{s,i}$).

where the relative frequency distribution of the i-th intervention is given by the ratio between the number of observations of this event and the number of total observations, that is the number of energy audits examined, coinciding with the number of companies constituting the sample (sample dimension—SD), see Equations (1) and (2):

$$f_{a,\,i} = \frac{n_{a,i}}{SD}, \tag{1}$$

$$f_{s,\,i} = \frac{n_{s,i}}{SD}, \tag{2}$$

The considered sample is only partially represented by Equations (1) and (2). Indeed, the studied cement plants can be classified into two separate categories: (i) half-cycle plant (i.e., grinding center) where only the finish grinding is performed and (ii) full-cycle plant where the entire manufacturing process is carried out. As a result, some interventions can be implemented for just a portion of the original sample. For instance, the interventions related to the kiln are applicable only to the full-cycle plants.

To produce a more realistic representation of the trends related to the adopted technologies in cement plant, Equations (3) and (4) are calculated for each intervention:

$$f_{relevant\_a,\,i} = \frac{n_{a,i}}{reference\_SD}, \tag{3}$$

$$f_{relevant\_s,\,i} = \frac{n_{s,i}}{reference\_SD}, \tag{4}$$

where $n_{a,i}$ and $n_{s,i}$ denote the number of applications and suggestions of a given intervention, respectively, while *reference_SD* represents the effective number of manufacturing sites that could implement the considered intervention. The application frequency of each solution is useful to assess its degree of diffusion in recent years. High values of $f_{relevant\_a,i}$ indicate that intervention was often applied in the last four years. On the other side, the frequency of suggestion represents the technology's degree of attractiveness for Italian companies in the current scenario [25]. High values of $f_{relevant\_s,\,i}$ indicate that the i-th solution is probable to be applied in the next four years. In other words, $f_{relevant\_a,i}$ keeps up to date with the actual level of application of energy efficiency measures in the Italian country, while $f_{relevant\_s,\,i}$ gives an idea of the future development perspectives of technologies in Italy.

After analyzing the data, by identifying the highest and the lowest $f_{relevant\_a,\,i}$ and $f_{relevant\_s,\,i}$, the most recently applied solutions and the most proposed ones were determined. In Section 3, the results obtained by applying this methodology are illustrated.

3. Results

From 65 energy audits, a sample of 48 Italian cement plants was selected. They are located in Italy, as shown in Figure 2. To obtain an adequate representation of the national scenario, companies were selected to represent both the regions in the north, the center, and the south of Italy. In a sample of 48 sites, 20 companies were in the norther regions of Italy (Emilia Romagna, Friuli Venezia Giulia, Lombardy, Piedmont, and Veneto), 8 in the center (Tuscany, Latium, and Umbria), and 20 in the south (Abruzzo, Basilicata, Calabria, Campania, Molise, Apulia, Sardinia, and Sicily).

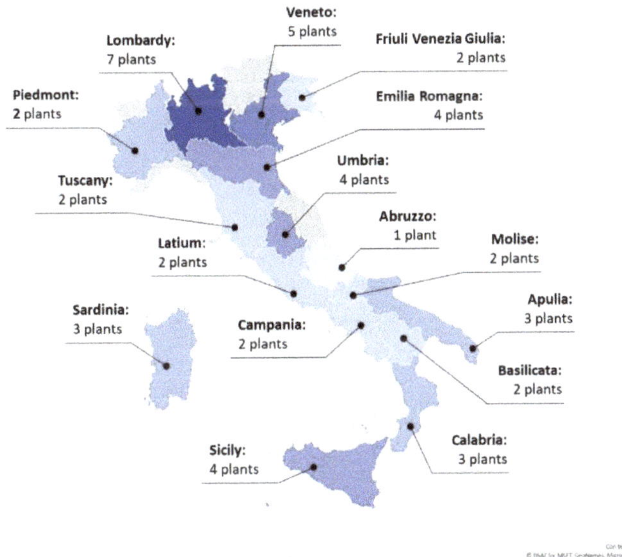

Figure 2. Geographical distribution of the 48 Italian companies constituting the sample.

In terms of manufacturing processes, 34 out of 48 sites carried out the complete cement production cycle, while 14 plants performed only the third stage of the manufacturing process (Figure 1, finish grinding process). Moreover, the finished products manufactured by the production sites were those shown in Table 1. The variety of production processes and finished products further confirms that this sample represents the national scenario well. In fact, plants producing cement and clinker (33.3% of the sample), plants producing only cement (33.3%), plants producing cement and other elements such as bituminous asphalt or other hydraulic binders (29.2%), and plants producing white cement (4.2%) were all taken into consideration.

Table 1. Types of finished products manufactured by the 48 production sites.

Type of Finished Products	Number of Production Sites	Percentage of the Sample
Cement and clinker	16	33.3%
Cement only	16	33.3%
Cement and other	14	29.2%
White cement	2	4.2%

The technological energy-saving solutions extrapolated from the energy audit and the results of the analysis are shown in Table 2. In Table 2, suggested and applied interventions are ordered according to process stages, putting all auxiliary and heat recovery solutions upstream because, as already explained, they are considered separately.

Table 2. Results of the analysis of Italian energy audits.

Process Stage	Process Machine	Solution Object	Intervention	$n_{a,i}$	$n_{s,i}$	$f_{a,i}$	$f_{s,i}$	$f_{relevant\_a,i}$	$f_{relevant\_s,i}$
Auxiliary systems	Engines	Engines	Installing efficient electric motors (class IE2, IE3, or IE4)	3	7	0.06	0.15	0.06	0.15
Auxiliary systems	Engines	Engines	Installing variable speed motors (motors with inverters)	1	8	0.02	0.17	0.02	0.17
Auxiliary systems	Pressure systems	Pressure systems	Installing inverters on compressors, pumps, or fans	4	19	0.08	0.40	0.08	0.40
Auxiliary systems	Pressure systems	Pressure systems	Insulating pipes, valves, and pumps and installing sealing to reduce air leakage	0	4	0	0.08	0	0.08
Auxiliary systems	Pressure systems	Silo cement extraction plant	Replacing pumps with rotocells	0	1	0	0.02	0	0.03
Auxiliary systems	Pressure systems	Chilled water distribution systems	Installing shut-off valves on the cooling water branches (pushed by pumps), to block flows when the system is at a standstill	0	1	0	0.02	0	0.03
Auxiliary systems	Electricity transformers	Electricity transformers	Optimizing transformer losses in the electrical cabin	0	1	0	0.02	0	0.02
Auxiliary systems	Electricity transformers	Electricity transformers	Replacing oil transformers with resin transformers (having less leakage)	0	2	0	0.04	0	0.04
Auxiliary systems	Electricity transformers	Electricity transformers	Renewing transformers in the electrical cabin installing k-factor transformers	0	1	0	0.02	0	0.02
Heat recovery systems	Heat recovery system	Heat recovery system	Installing an ORC turbine for electricity production	2	4	0.04	0.08	0.06	0.12
Heat recovery systems	Heat recovery system	Kiln	Installing heat recovery systems to use the hot gases leaving the kiln to dry raw materials	0	2	0	0.04	0	0.06
Heat recovery systems	Heat recovery systems	Kiln	Installing heat exchanger to recover heat from the flue gases and pre-heat the thermal oil for the kiln fuel	1	0	0.02	0	0.03	0
Heat recovery systems	Heat recovery systems	Cooler	Installing systems to recover heat from the thermal waste of the cooler (useful for heating offices or other)	0	2	0	0.04	0	0.06
Crushing	Crusher	Feeder	Installing modern gravimetric feeders and scales	1	0	0.02	0	0.03	0
Grinding raw materials	Mill	Mill	Installing a high-pressure roller mill	0	1	0	0.02	0	0.03
Grinding raw materials	Mill	Mill/Separator	Installing a mill with a dynamic separator and cyclones with a process filter	1	0	0.02	0	0.03	0
Grinding raw materials	Mill	Boilers	Replacing boilers with more advanced and efficient ones	1	0	0.02	0	0.03	0
Grinding raw materials	Mill and cyclones	Mill and cyclones	Eliminating two cyclones after introducing an electrofilter and a bag filter for the mill	1	0	0.02	0	0.03	0
Blending	Homogenizing silo	Silo's blowers	Replacing silo's blowers with inverter-equipped screw compressors	0	1	0	0.02	0	0.03
Preheating	Preheating tower	Cyclones	Reducing the cyclones from 16 to 12 in the Lepol grid preheater	1	0	0.02	0	0.03	0
Heating	Kiln	Kiln	Installing an automatic conduction system for the kiln	0	2	0	0.04	0	0.06
Heating	Kiln	Burners	Install an advanced dosing device for dosing powder to the main burner of the kiln to optimize consumption	1	0	0.02	0	0.03	0

Table 2. *Cont.*

Process Stage	Process Machine	Solution Object	Intervention	$n_{a,i}$	$n_{s,i}$	$f_{a,i}$	$f_{s,i}$	$f_{relevant\_a,i}$	$f_{relevant\_s,i}$
Heating	Kiln	Burners	Installing intermediate inverter exhausters for the kiln to optimize the power output during the firing process	1	0	0.02	0	0.03	0
Final milling	Mill	Feeder	Optimizing the feeding system installing more efficient and advanced systems	0	2	0	0.04	0	0.04
Final milling	Mill	Dispenser	Installing weight-measuring devices for material entering the mill	1	0	0.02	0	0	0.02
Final milling	Mill	Mill	Installing an automatic conduction system for ball mills	0	1	0	0.02	0	0.02
Dedusting	Dust filtration system	Dust filtration system	Installing an electrofilter	1	0	0.02	0	0.03	0
		Total		20	59	-	-	-	-

Table 2 was created after analyzing energy audits and also after screening the scientific literature on technological solutions to reduce energy consumption in the cement industry. The content of Table 2 is characterized by two colors. In black are outlined the solutions suggested or applied by the sample of companies and previously mentioned by other scientific authors (bibliographic references and detailed explanations are given in Appendix A). In red, instead, are represented the solutions identified in the energy audits, but not found in the scientific literature.

Figure 3 shows the applied and suggested interventions by considering the process stage on which they acted to reduce energy consumption, whereas Figure 4 summarizes the same measures according to the process machinery (see Figure 1).

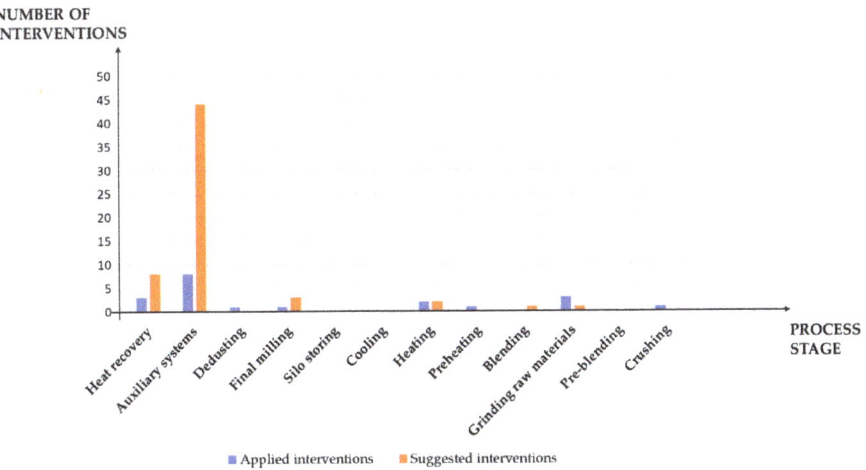

Figure 3. Suggested and applied interventions divided according to the process stage on which they act.

Results in Table 2 show a strong tendency of analyzed companies to prefer improvements in auxiliary processes, working on engines, compressors, fans, pumps, or on heat recovery systems over all the other interventions reported in Tables A1–A17 (Appendix A). The aforementioned improvements are preferred to other interventions, due to the ease of implementation and the low operative costs required. In contrast, no solution is adopted or suggested for the pre-blending and silo storing phases. Neither are considered solutions to improve the sustainability of equipment such as stackers and reclaimers.

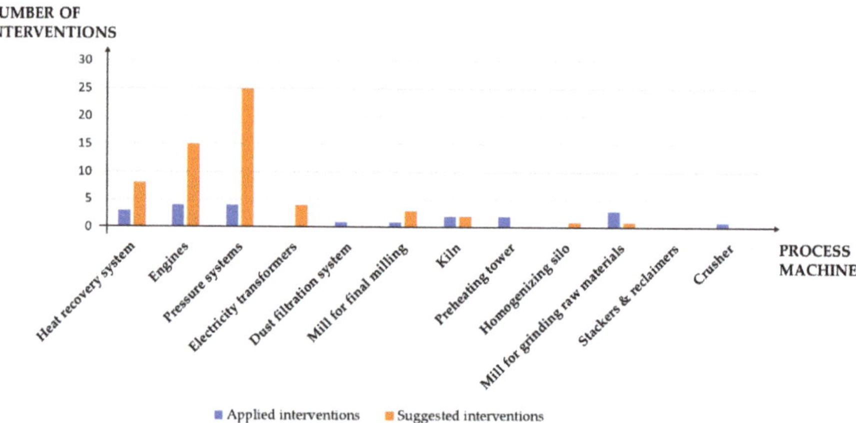

Figure 4. Suggested and applied interventions divided according to the process machine on which they act.

3.1. Analysis of the Applied Interventions

The sum of column $n_{a,i}$ (4th column of Table 2) is not 48 (sample dimension) but assumes a value of 20. This is because some companies carried out more than one intervention, while others did not apply any improvements in the last four years.

In the case of recently applied interventions, most of the sample companies acted on the auxiliary systems, for instance installing variable speed motors, pumps, or compressors, or replacing boilers and motors with more efficient ones.

Another solution implemented by 2 out of 34 production sites was to replace the classic steam cycle with the Organic Rankine Cycle (ORC) to recover heat from hot gases and produce electricity. Today, however, this solution does not represent the market standard in Italy, especially due to the high investment costs involved in its implementation (as confirmed by the analyses shown below and summarized in Tables 3–5).

Table 3. Energy savings produced by the applied technological measures in the various areas of intervention. The total annual savings are calculated as the sum of thermal energy, electricity, and fuel savings.

Area of Intervention	Number of Production Sites Reporting Quantitative Information	Electricity Savings (Toe/Year)	Thermal Energy Savings (Toe/Year)	Fuel Savings (Toe/Year)	Annual Savings (Toe/Year)	Annual Savings (%)	Average Annual Savings (Toe/Year)
Pressure systems	2	96	66	0	162	2.6%	81
Thermal power plant and heat recovery systems	5	1844	139	140	2123	33.4%	425
Engines, inverters, and other electrical installations	2	27	0	0	27	0.4%	27
Production lines and machines	4	87	3952	0	4039	63.6%	1010
Total	13	2054	4157	140	6351	100%	-

Overall, the most implemented technological solution over the last four years was the installation of inverters on compressors, pumps, or fans. Such a solution showed a relative frequency distribution ($f_{relevant\_a,i}$) of 0.08.

Some companies also reported quantitative data on the savings achieved by implementing technological measures. These results are divided by area of intervention and summarized in Tables 3 and 4. In the tables, toe stands for ton of oil equivalent. Areas related to thermal recovery and production lines determine large energy savings, and the largest economic investments (both total and average). The average quantitative data shown in Tables 3 and 4 is computed as the average of the number of production sites that reported quantitative information.

Table 4. Investments required to apply technological measures in the various areas of intervention.

Area of Intervention	Number of Production Sites Reporting Quantitative Information	Total Investment (€)	Total Investment (%)	Average Investment (€)
Pressure systems	2	124,798	0.4%	62,399
Thermal power plant and heat recovery systems	3	16,220,000	58.4%	5,406,667
Production lines and machines	4	11,430,000	41.2%	2,857,500
Total	9	27,774,798	100%	-

Table 5. Cost-effectiveness indicator for each area of intervention.

Area of Intervention	Number of Production Sites Reporting Quantitative Information	Cost-Effectiveness Indicator (€/toe)
Pressure systems	2	874
Thermal power plant and heat recovery systems	3	16,076
Production lines and machines	4	10,940

A cost-effectiveness indicator was calculated for each intervention, measured as Euros invested per Ton of Oil Equivalent (toe) of energy saved (Table 5). The available information allowed to calculate it only on nine interventions, reporting both information on energy saved and costs. The area of pressure system interventions shows a particularly advantageous indicator, thus explaining the current tendency of Italian companies to invest in auxiliary systems.

3.2. Analysis of the Suggested Interventions

On the contrary, the sum of column $n_{s,i}$ (5th column of Table 2) is greater than 48 (sample size), assuming a value of 59. The significant difference between the total of $n_{s,i}$ (59) and the total of $n_{a,i}$ (20) may suggest an interest of Italian companies to improve their sustainability in the next four years.

Even in the case of the proposed interventions, most of the companies showed interest in technological solutions related to auxiliary systems, proposing to exploit variable speed machines, to improve the energy class of engines, or to improve the electricity transformers. Another suggested solution was to use the ORC cycle to produce electricity (4 out of 34 companies).

Once again, the most attractive and most proposed technological solution for the next four years was the installation of inverters on compressors, pumps, or fans. Such a solution showed a relative frequency distribution ($f_{relevant\_s,i}$) of 0.40.

Tables 6 and 7 summarize the energy savings and investment cost indicated by those companies that proposed a feasibility study. Table 8 reports the cost-effectiveness indicators calculated for the suggested interventions. Feasibility studies estimated electrical savings to be far greater than thermal savings in all areas, except the production lines and machines one. This result was strongly influenced by the technological measures in heat recovery systems. As in the applied interventions, also in the suggested measures, the highest energy saving was associated to the heat recovery area, accompanied however by a significant investment cost (Table 7). This area shows a relatively good cost-effectiveness indicator; indeed, only production lines and machines area has a better value of cost-effectiveness than thermal power plant and heat recovery systems (Table 8).

Table 6. Energy savings assessed for the suggested technological measures in the various areas of intervention. The total annual savings are calculated as the sum of thermal energy, electricity, and fuel savings.

Area of Intervention	Number of Production Sites Reporting Quantitative Information	Annual Electricity Savings (Toe/Year)	Annual Thermal Energy Savings (Toe/Year)	Annual Savings (Toe/Year)	Annual Savings (%)	Average Annual Savings (Toe/Year)
Pressure systems	18	446	0	446	2.2%	25
Thermal power plant and heat recovery systems	10	12,546	29	12,575	63.4%	1258
Engines, inverters, and other electrical installations	15	998	0	998	5.0%	263
Production lines and machines	20	526	5296	5821	29.3%	306
Total	63	14,516	5325	19,840	100%	-

Table 7. Investments assessed for the suggested measures in the areas of intervention.

Area of Intervention	Number of Production Sites Reporting Quantitative Information	Total Investment (€)	Total Investment (%)	Average Investment (€)
Pressure systems	18	1,392,025	3.0%	77,335
Thermal power plant and heat recovery systems	4	34,540,000	73.8%	8,635,000
Engines, inverters, and other electrical installations	15	6,945,793	14.8%	1,817,566
Production lines and machines	18	3,941,626	8.4%	218,979
Total	55	46,891,444	100%	-

Table 8. Cost-effectiveness indicator for each area of intervention.

Area of Intervention	Number of Production Sites Reporting Quantitative Information	Cost-Effectiveness Indicator (€/toe)
Pressure systems	17	4491
Thermal power plant and heat recovery systems	4	3788
Engines, inverters, and other electrical installations	15	7158
Production lines and machines	16	2649

Suggested technological measures can also be analyzed distinguishing for their payback time class (PBT, Figure 5). In this case, 55 measures report quantitative information: measures with PBT between one and two years represent 40% (5.3 ktoe/year) of total annual potential saving. Further 20% of potential saving is associated with measures having a PBT between 3 and 5 years (3.2 ktoe/year).

Finally, Figure 6 shows that 75% of potential saving (10.6 ktoe/year) can be achieved by mobilizing 40% of total investment associated with suggested measures (around 19 million Euro), highlighting that relatively less expensive measures are associated with a high saving potential.

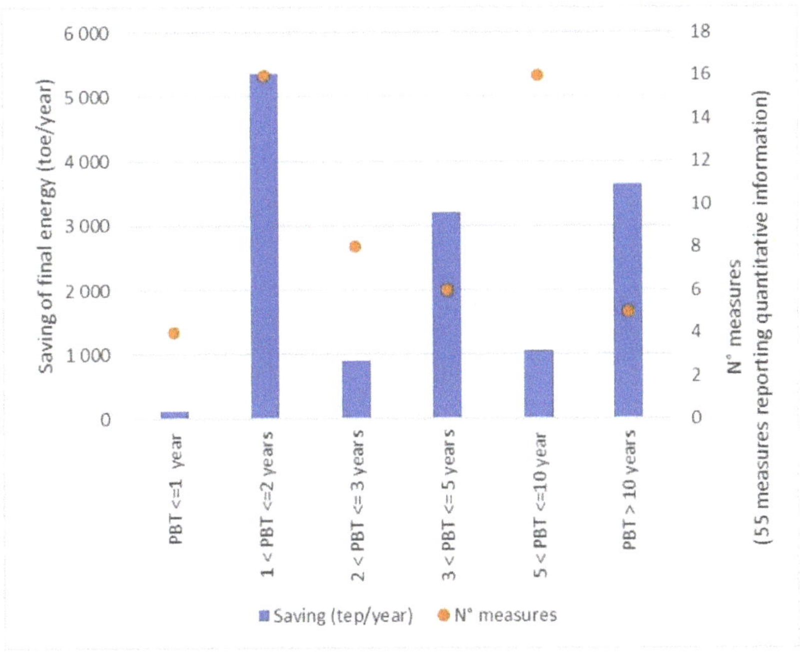

Figure 5. Annual saving and suggested measures according to PBT classes.

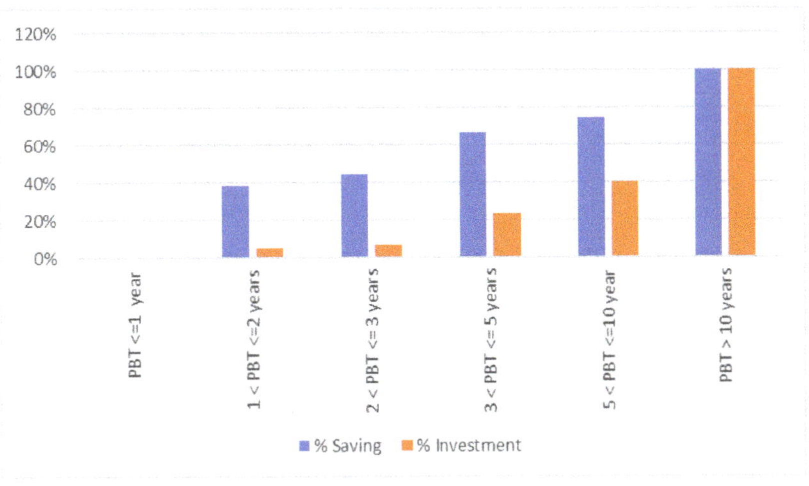

Figure 6. Cumulative saving and investment according to PBT classes.

3.3. Other Results

In the energy audits, the sampled companies mentioned some technological interventions not found in the literature (red writings, Table 2). Among them, there are improvements in the electricity transformers, several interventions to make the kiln more efficient, some solutions to increase the sustainability of final milling, and so on. The other solutions listed in Table 2 were already suggested by scientific authors [4,15–17,34,35]. In this case, by mentioning them, expert industrialists confirmed their current goodness as sustainable technologies. As anticipated in Section 2, the results of the analyses were shared with expert industrialists to validate the generated output. Overall, consulting energy audits and sector experts allowed validating 10 solutions from the literature, while expanding Tables A1–A17 (Appendix A) by adding 17 solutions. More information on the latter can be found in Appendix A.

4. Discussion

Today, sustainability is an aspect that is becoming increasingly important in industrial plants. It has many facets and covers aspects ranging from the location of production facilities [36], to the optimization of logistics in industrial plants [37], to the efficient energy consumption in production systems [1]. In terms of energy consumption, one of the most energy-intensive processes is cement production. Indeed, the cement industry is highly energy-intensive, determining around 7% of global industrial energy consumption each year.

Improving production technologies, by replacing inefficient equipment with better-performing ones, is a good strategy to reduce the energy consumption of a cement plant. In this sense, the market offers a wide range of solutions, and the scientific literature already provides several reviews of available energy-saving technologies [38].

However, technology is constantly developing, so the alternatives may change within a few years. Some solutions may become obsolete and fall into disuse while others may gain interest due to technological progress. For this reason, as shown by [19,25], besides consulting the literature, a viable way to keep the knowledge updated and also investigate the current attractiveness of each solution is to analyze a sample of real companies. To the best of the authors' knowledge, similar work is missing for Italian companies, although Italian cement plants are among the most significant in Europe.

After a preliminary consultation of the existing literature to identify the available technological energy-saving solutions, based on the approach by Hasanbeigi et al. [27], in this paper, this gap is filled by investigating a sample of 48 production sites, analyzing production and energy consumption data (in a normalized and anonymous way) from energy audits collected by ENEA, pursuant to Art. 8 of the EED Directive.

In this paper, by applying the approach described in Section 2, the following information is investigated in energy audits: technological energy efficiency measures implemented by the companies in the last four years, sustainable energy efficiency measures proposed to be implemented in the next four years, the relative frequency distribution of application and the relative frequency distribution of suggestion of each technology, and, finally, the motivations that led each company to be interested in the respective proposed solutions. The degree of application and future perspectives of available technologies to reduce energy consumption are studied considering as a reference sample size the number of plants where a given intervention can be implemented. This information answers the research questions RQ1 and RQ2.

The main reasons for the choices and trends in technologies were also researched, analyzing the costs of the solutions, the energy benefits they bring, and, above all, the cost-effectiveness factor, given by the ratio between the cost of solutions and their impact on consumptions. This last aspect answers research question RQ3.

Results prove that in Italy, companies have focused their attention mainly on solutions to reduce the consumption of auxiliary systems such as compressors, engines, pumps, and fans. This is because of their easiness of implementation and the low operative costs required. This claim is validated by the historical data and feasibility studies reported by some of the surveyed companies.

In particular, the most attractive (most applied and most suggested) solution was the installation of inverters on compressors, pumps, or fans. Such a solution showed the following relative frequency distributions: $f_{relevant\_a} = 0.08$ and $f_{relevant\_s} = 0.40$.

Therefore, based on the results, it is possible to say that Italian companies will likely invest in technologies with a high relative frequency distribution value. Inefficient, non-variable speed engines and turbomachines will probably become obsolete, and a spread of

ORC cycle systems for heat recovery is expected to occur (compatibly with budget availability and the eventual possibility of exploiting the opportunities of outsourcing the energy service).

Overall, the results of the analysis do not allow for the calculation of the payback period (PBP) of each technological intervention. However, by consulting the cost-effectiveness indicators and the applied and suggested solutions, we can infer that PBP drives the choice of companies. PBP is likely a parameter proportional to frequency, and low PBP values make technological solutions preferable.

It is worth mentioning the relevance of the effective sample size to estimate the frequencies of application and suggestion, indeed considering all the manufacturing plants for the interventions that can be implemented only for the full-cycle sites, would underestimate the frequencies. On the contrary, less appreciation will have to improve the pre-blending and silo storing processes or make stackers and reclaimers efficient.

Besides forecasting the development of technologies in Italy, the consultation of expert industrialists allowed collecting updated ideas for improving technologies, thus giving valuable inputs to scientific research. To the 198 solutions previously described by other scientific authors (Tables A1–A17, Appendix A, black writings), 17 solutions were added (Tables A1–A17, Appendix A, red writings). With regards to the final milling phase, it emerged that companies prefer making the upstream feeding more efficient by acting on feeders and dispensers, while literature insists on modifying the mill. Moreover, companies proposed solutions on transformers or auxiliary systems, whose opportunities are not emphasized in the cement sector literature. Furthermore, both literature and companies have shown interest in upgrading kilns. Indeed, in addition to the 18 solutions already expressed by scientific authors, 3 others were added by consulting energy audits, mainly concerning the improvement of burners. Finally, consulting audits allowed us to identify further solutions to the ones already found in the literature to make the preheating cyclones and blending silo more sustainable (see Appendix A).

5. Conclusions

The presented analysis of energy audits was useful to identify the degree of application and the future perspectives of the available technologies in the Italian industry, also allowing to identify the main reasons for these choices and trends. The outcome of the analysis is a realistic representation of the current Italian scenario, which clearly suggests that the choices of Italian companies are driven by considerations concerning the Pay Back Period of technological solutions. Indeed, companies currently prefer to act on auxiliary systems, improving the efficiency of engines and pressure systems and, especially, adopting variable speed drives. The low value of the cost-effectiveness indicator explains why companies have invested in such technological solutions in the last four years and why they will presumably continue doing so in the near future. Another noteworthy result of this study can be summarized as follows. Despite a high investment cost, companies are showing increasing interest in ORC cycle systems for heat recovery, and a spread of these solutions is expected to occur soon.

In addition to the description of the Italian scenario, a collateral outcome of this paper is an updated list of energy-saving technologies (which are divided by process stage and machine type). This work was entirely developed in cooperation with Federbeton. In particular, the analysis of the audits and the creation of the list of technologies (Appendix A) was carried out by the University of Florence and ENEA. Subsequently, this list was shared with Federbeton and the observations of plant managers and other sector experts were collected. Consequently, the generated list of technologies was validated by both the main Italian entities in the field of cement (Federbeton) and energy (ENEA) efficiency. The provided list of technologies could be useful for companies willing to improve their sustainability by implementing new technological opportunities. In fact, a concise and up-to-date representation of possible strategies to reduce the consumption of technologies can help industrialists to clarify what maneuvers can be implemented in the various plant areas, both suggested by literature studies and sector experts. In this sense, the tables provided in Appendix A can represent a decision support tool.

There are two limitations in this research: First, to keep the list of technological measures (Appendix A) updated, the analysis here proposed should be regularly repeated in the future. Second, the present study is mainly based on the consultation of energy audits carried out by obliged companies, which are only a part of all Italian companies in the cement sector.

However, the results emerging from the analysis could be interesting both at the academic and industrial levels. In fact, consulting real companies allows obtaining a precise view of the currently interesting technologies in the specific production sector considered, thus updating the state of the art of literature. Moreover, a generic company willing to improve its sustainability in a specific plant area/machine (for instance the one affected by higher energy consumption) can consult the present paper to quickly identify the available alternatives. In this perspective, this article not only gives a clear picture of the Italian situation but could also be useful for companies in the cement sector working outside the Italian nation.

Future developments of this work could be as follows. First of all, it would be interesting to develop a multi-criteria ranking of the available technological solutions, evaluating aspects such as installation cost, operational and maintenance costs, payback period, energy-saving achieved, and so on. The obtained ranking could help companies in choosing the most suitable solution based on their requirements and needs. Secondly, it would be worthwhile studying other countries, to verify whether the Italian situation is in line with the international trend. Thirdly, the proposed study could be repeated in manufacturing sectors other than cement. Finally, the proposed study could be repeated in the future (say in the next ten years) to determine if radical changes have happened or not.

Author Contributions: Conceptualization, A.C., L.L., and F.D.C.; methodology, A.C., L.L., and F.D.C.; software, C.M., M.S., F.M., A.C., and L.L.; validation, A.C., L.L., F.D.C., F.M., M.S., and C.M.; formal analysis, A.C., L.L., and C.M.; investigation, A.C., L.L., and C.M.; resources, C.M., M.S.; data curation, A.C., L.L., and C.M.; writing—original draft preparation, A.C.; writing—review and editing L.L., F.D.C., C.M., F.M., and M.S.; visualization, A.C.; supervision, F.D.C.; project administration, F.D.C., F.M., and M.S.; funding acquisition, M.S. and F.M. All authors have read and agreed to the published version of the manuscript.

Funding: This work is part of the Electrical System Research (PTR 2019-2021), implemented under Programme Agreements between the Italian Ministry for Economic Development and ENEA, CNR, and RSE S.p.A.

Institutional Review Board Statement: Not applicable.

Informed Consent Statement: Not applicable.

Data Availability Statement: Not applicable.

Acknowledgments: The authors gratefully acknowledge the technical support provided by the Italian Association Federbeton Confindustria.

Conflicts of Interest: The authors declare no conflict of interest.

Appendix A

The principal technological solutions described in the literature for reducing the energy consumption of a cement production plant can be summarized as shown in Tables A1–A17. Each table refers to the strategies (3rd column) that can be implemented to improve a specific process operation (table header), dividing the available alternatives according to the machinery (1st column) and the object of the machinery involved (2nd column).

Each technological alternative is associated with the bibliographic reference consulted to identify it (4th column). In light of this, readers interested in finding out more about individual solutions can consult the 4th column of each Table.

To create Tables A1–A17, we initially followed the approach proposed by [24], identifying the major production processes and corresponding energy efficient technologies by screening the previous literature studies. The solutions identified during this stage were written in black in the tables.

Subsequently, as already described in Section 2, other solutions detected through the energy audits were added to the alternatives found in the literature. The solutions identified in this phase were written in red in the tables. The list of technology solutions was shared with Federbeton and its content was reviewed and validated by sector experts. Only for the solutions indicated by the experts as currently obsolete, appropriate comments were inserted in a 5th column.

Table A1. Technological energy-saving solutions for crushing.

Process Machinery	Solution Object	Energy-Saving Technological Solution	Reference
Crusher	Feeder	Installing modern gravimetric feeders and scales for efficient raw material feeding into the crusher	[4]
Crusher	Screening	Installing machines for preliminary screening	[4]
Crusher	Screening	Improving screening efficiency through efficient designs	[10]
Crusher	Crusher	Installing a hammer crusher (preferable if the moisture content is less than 10%) with or without a screen	[4]
Crusher	Crusher	Installing a twin-rotor hammer crusher (preferable if the moisture content less than 10%) with or without a screen	[39]
Crusher	Crusher	Installing a compact impact crusher (also known as a monorail impact crusher)	[39,40]
Crusher	Crusher	Installing an impact crusher (also known as a twin-rotor impact crusher)	[4]
Crusher	Crusher	Installing a jaw crusher (also known as a toggle crusher)	[4]
Crusher	Crusher	Installing a double toggle crusher	[39]
Crusher	Crusher	Installing a rotary crusher	[4]
Crusher	Crusher	Installing a roller crusher (cylinders)	[4]
Crusher	Crusher	Installing a toothed roller crusher (teeth ensure better distribution of the forces involved)	[39]
Crusher	Crusher	Installing a single toothed roller crusher	[39]
Crusher	Crusher	Installing a semi-mobile or mobile crusher	[4]

Table A2. Technological energy-saving solutions for quality control.

Quality Control			
Process Machinery	**Solution Object**	**Energy-Saving Technological Solution**	**Reference**
X-ray laboratory	X-ray laboratory	Installing an XRF laboratory (X-ray fluorescence)	[20]
X-ray laboratory	X-ray laboratory	Installing an automated XRF laboratory (X-ray fluorescence)	[20]
X-ray laboratory	X-ray laboratory	Installing a laboratory with a robotic and automated EDXRF spectrometer to sample the material mixture entering the mill	[20]
X-ray laboratory	X-ray spectrometer	Installing a cross-belt XRF EDXRF spectrometer on the conveyor belt from the pre-homogenization site to the mill	[20]
γ-ray laboratory	γ-ray laboratory	Installing a laboratory with a gamma-ray spectrometer (batch γ-ray)	[20]
γ-ray laboratory	γ-ray spectrometer	Installing a gamma-ray spectrometer at the crusher outlet (chute γ-ray)	[20]
γ-ray laboratory	γ-ray spectrometer	Installing a cross-belt γ-ray spectrometer on the conveyor belt from the pre-homogenization site to the mill	[20]
γ-ray laboratory	γ-ray spectrometer	Installing a gamma-ray spectrometer at the crusher outlet (slurry γ-ray)	[20]

Table A3. Technological energy-saving solutions for pre-blending.

Pre-Blending			
Process Machinery	**Solution Object**	**Energy-Saving Technological Solution**	**Reference**
Stacker	Stacker	Installing a circular stacker	[40]
Stacker	Stacker	Installing a longitudinal stacker	[40]
Reclaimer	Reclaimer	Installing a circular reclaimer	[40]
Reclaimer	Reclaimer	Installing a longitudinal reclaimer	[11,40]

Table A4. Technological energy-saving solutions for slurry production.

Slurry Production				
Process Machinery	**Solution Object**	**Energy-Saving Technological Solution**	**Reference**	**Comments from Sector Experts**
Wash-mill	Wash-mill	Installing a closed-circuit washer	[4,27]	
Mill	Mill	Installing a tubular mill for wet and semi-wet processes	[4]	Solution reported in literature, but currently obsolete
Mill	Mill	Installing a ball mill for wet and semi-wet processes	[4]	Solution reported in literature, but currently obsolete

Table A5. Technological energy-saving solutions for drying.

Drying			
Process Machinery	**Solution Object**	**Energy-Saving Technological Solution**	**Reference**
Dryer	Dryer	Installing a rotary drum dryer with parallel flow	[41]
Dryer	Dryer	Installing a rotary drum dryer with opposing flows	[41]
Dryer	Dryer	Installing a solar dryer	[12]
Dryer	Dryer	Installing a rapid dryer	[40]
Dryer	Dryer	Installing a pre-dryer before a ball mill or tube mill	[40]
Dryer	Dryer	Installing a filter press (for the semi-wet process)	[9]
Dryer	Dryer	Installing a dryer-pulverizer (dedicated machine for simultaneous crushing and drying)	[40]
Dryer	Dryer	Installing an impact dryer	[40]
Dryer	Dryer	Installing a tandem drying grinding (a combination of hammer mill and ventilated ball mill)	[40]
Dryer	Dryer	Install mechanical air separators	[40]

Table A6. Technological energy-saving solutions for grinding raw materials.

Grinding Raw Materials			
Process Machinery	**Solution Object**	**Energy-Saving Technological Solution**	**Reference**
Mill	Transport systems	Installing a weighing belt for raw materials entering the mill	[4]
Mill	Mill	Installing a tube mill with central discharge	[4]
Mill	Mill	Installing a tube mill with a closed-circuit final discharge	[4]
Mill	Mill	Installing a ball mill	[4]
Mill	Mill	Installing a ventilated ball mill	[4]
Mill	Mill	Installing a central discharge ball mill	[4]
Mill	Mill	Installing a ball mill with a closed-circuit final discharge	[4]
Mill	Mill	Installing a horizontal roller mill	[4,15,27]
Mill	Mill	Installing an autogenous mill	[4]
Mill	Mill	Installing a high-pressure roller mill	[4]

Table A6. *Cont.*

Grinding Raw Materials			
Process Machinery	**Solution Object**	**Energy-Saving Technological Solution**	**Reference**
Mill	Mill	Installing a track and ball vertical mill	[39]
Mill	Mill	Installing a vertical roller mill	[13,16,35]
Mill	Mill	Grinding the various materials entering the mill separately according to their fineness	[35]
Mill	Mill	Replacing and improving the abrasive material used for the balls	[16]
Mill	Mill	Installing a roller press in addition to a ball mill	[16,35]
Mill	Mill	Grinding the various materials entering the mill separately according to their hardness	[35]
Mill	Mill	Installing a dopplerotator mill	[42]
Mill	Mill	Installing an aerofall-mill (autogenous mill)	[40]
Mill	Mill	Installing a ventilated mill	[40]
Mill	Mill/Separator	Installing a mill with a dynamic separator and cyclones with a process filter	Energy audits
Separator	Separator	Installing a high-efficiency rotating cage separator	[4,9,35]
Separator	Separator	Installing a high-efficiency air separator	[4,9,23]
Separator	Separator	Installing a static grid separator	[40]
Mill	Mill's fans	Installing a three-fan system and one fan that is charged to the operations of the vertical roller crusher (cylinders)	[15]
Mill	Boilers	Replacing boilers with more advanced and efficient ones	Energy audits
Mill and cyclones	Mill and cyclones	Eliminating two cyclones after introducing an electrofilter and a bag filter for the mill	Sample of analyzed companies

Table A7. Technological energy-saving solutions for blending.

Blending			
Process Machinery	**Solution Object**	**Energy-Saving Technological Solution**	**Reference**
Blending silo	Blending silo	Removing leaks in the compressed air circuit by installing seals or other devices	[9,14]
Blending silo	Blending silo	Sizing the air circuit correctly	[9,14]
Blending silo	Blending silo	Installing a flow-controlled, multi-outlet blending silo	[40]
Blending silo	Blending silo	Installing a flow-controlled cone blending silo	[40]
Blending silo	Blending silo	Installing a turbulence blending silo	[40]
Blending silo	Blending silo	Restoring the structural integrity of the blending silo and eliminate leaks in its structure	[14]
Blending silo	Blending silo	Installing a gravitational blending silo	[13,16,27]
Blending silo	Blending tank	Installing a blending tank	[9]
Blending silo	Tank's agitator	Installing high-efficiency agitators	[9]
Blending silo	Silo's blower	Replacing silo's blowers with inverter-equipped screw compressors	Energy audits

Table A8. Technological energy-saving solutions for preheating.

Preheating				
Process Machinery	**Solution Object**	**Energy-Saving Technological Solution**	**Reference**	**Comments from Sector Experts**
Preheater	Preheater	Installing a cyclone preheater with calciner	[4]	
Preheater	Preheater	Installing a grid preheater with calciner	[4]	Solution reported in literature, but currently obsolete
Preheater	Preheater	Installing a cyclone preheater	[4]	
Preheater	Preheater	Installing a grid preheater	[4]	Solution reported in literature, but currently obsolete
Preheater	Preheater	Recovering hot gases from the preheating tower	[4]	
Preheater	Cyclones	Increasing the number of cyclones in the cyclone preheater	[15,27,35]	
Preheater	Cyclones	Optimizing the number of cyclone stages considering the characteristics and properties of the raw materials and fuels used	[4]	
Preheater	Cyclones	Reducing the cyclones from 16 to 12 in the Lepol grid preheater	Sample of analyzed companies	
Preheater	Cyclones	Replacing the cyclone preheater with a low pressure drop cyclone preheater	[13,15,17,23,27,35]	
Preheater	Preheater's fans	Installing high-efficiency fans in the preheater	[9,15,27]	
Pre-calciner	Pre-calciner	Replacing the existing pre-calciner with a more advanced and efficient one	[13]	
Pre-calciner	Pre-calciner	Installing a pre-calciner with a high solid/gas ratio	[4,18]	
Pre-calciner	Burners	Replacing calciner burners	[4]	

Table A9. Technological energy-saving solutions for heating.

	Heating			
Process Machinery	Solution Object	Energy-Saving Technological Solution	Reference	Comments from Sector Experts
Kiln	Dispensers	Installing gravimetric dosing units for efficient flour feeding into the kiln	[4]	
Kiln	Kiln	Installing a rotary kiln	[4]	
Kiln	Kiln	Installing a vertical kiln	[4]	
Kiln	Kiln	Improving the refractory lining of the kiln	[9,15,17,23,27]	
Kiln	Kiln	Installing a fluidized bed advanced cement kiln	[18]	Emerging solution
Kiln	Kiln	Improving the combustion system	[4,18]	
Kiln	Kiln	Minimizing leakage or entry of unwanted air/gases into the kiln. Alternatively, replacing inlet and outlet seals	[9,17]	
Kiln	Kiln	Stabilizing the outer shell of the kiln	[4,18]	
Kiln	Kiln	Minimizing bypass flow	[4,18]	
Kiln	Kiln	Installing indirect combustion systems	[4,18]	
Kiln	Kiln	Reducing the pressure of the methane gas coming from international network through a turbine. The turbine can be connected to a generator to produce electrical energy.	[4]	
Kiln	Kiln	Replacing the existing kiln with a higher capacity one	[13,35]	
Kiln	Kiln	Installing an automatic conduction system for the kiln	Energy audits	
Kiln	Kiln's fans	Installing fans for cooling the kiln with a larger inlet diameter	[9]	
Kiln	Fuel transport system	Installing a gravimetric feeding system to efficiently feed solid fuel	[4]	
Kiln	Fuel transport system	Installing pumps for liquid fuel burners	[4]	
Kiln	Burners	Installing single-channel burners	[4]	
Kiln	Burners	Installing multi- channel burners	[4]	
Kiln	Burners	Installing an advanced dosing device for dosing powder to the main burner of the kiln to optimize consumption	Energy audits	
Kiln	Burners	Installing intermediate inverter exhausters for the kiln to optimize the power output during the firing process	Energy audits	

Table A10. Technological energy-saving solutions for cooling.

Cooling				
Process Machinery	**Solution Object**	**Energy-Saving Technological Solution**	**Reference**	**Comments from Sector Experts**
Cooler	Cooler	Installing a tube cooler	[4]	Solution reported in literature, but currently obsolete
Cooler	Cooler	Installing a planetary cooler	[4]	Solution reported in literature, but currently obsolete
Cooler	Cooler	Installing a travelling grid cooler	[4]	
Cooler	Cooler	Installing an oscillating grid cooler	[4]	
Cooler	Cooler	Replacing the ventilation system of the touring cooler with a more efficient one	[4,13,35]	
Cooler	Cooler	Adding a static grid to the touring grid cooler	[4,13,35]	
Cooler	Cooler	Installing a vertical cooler	[4]	
Cooler	Cooler	Replacing existing grid coolers with high-efficiency ones	[4]	
Cooler	Cooler	Optimizing energy recovery from the cooler by means of a static grid	[4]	
Cooler	Cooler	Installing a rotating cooling disc	[4]	
Cooler	Cooler	Installing a rapid cooler (only for white cement)	[4]	
Cooler	Cooler	Installing a pendulum frame for swinging grilles	[11]	
Cooler	Grid	Replacing existing grilles with more advanced and modern ones	[4]	
Cooler	Grid	Replacing the grill plates of the second-generation touring cooler	[4,13,35]	

Table A11. Technological energy-saving solutions for final milling.

Final Milling				
Process Machinery	**Solution Object**	**Energy-Saving Technological Solution**	**Reference**	**Comments from Sector Experts**
Mill	Dispensers	Installing weight-measuring devices for material entering the mill	Energy audits	
Mill	Feeder	Optimizing the feeding system installing more efficient and advanced systems	Energy audits	
Mill	Feeder	Installing a weighing belt for each material entering the mill	[4]	
Mill	Feeder	Installing a clinker flow regulator at the tube mill inlet	[4,18]	
Mill	Mill	Installing a ball mill with a closed circuit	[4]	
Mill	Mill	Installing a ball mill with closed circuit final discharge	[4]	
Mill	Mill	Installing a ball mill with open circuit final discharge	[4]	Solution reported in literature, but currently obsolete
Mill	Mill	Installing a vertical roller mill	[13]	
Mill	Mill	Installing a horizontal roller mill	[4]	
Mill	Mill	Installing a high-pressure roller press	[4]	
Mill	Mill	Installing a vertical roller mill before the ball mill	[4,18]	
Mill	Mill	Installing a roller press before the ball mill	[4,18]	Uncommon solution
Mill	Mill	Improving the abrasive material used for the balls or replacing the balls with better performing ones	[4,18]	
Mill	Mill	Installing a classification liner for the ball mill at the entrance to the second chamber	[4,18]	
Mill	Mill	Installing plasma technology	[18]	Emerging solution
Mill	Mill	Installing ultrasound technology	[18]	Emerging solution
Mill	Mill	Installing an automatic conduction system for ball mills	Energy audits	
Mill	Mill's fans	Replacing mill fans with high-efficiency ones	[15]	

Table A12. Technological energy-saving solutions for fuel grinding.

Fuel Grinding			
Process Machinery	**Solution Object**	**Energy-Saving Technological Solution**	**Reference**
Mill	Fuel transport system	Installing a weighing belt for raw materials entering the mill	[4]
Mill	Mill	Installing a ventilated ball mill for solid fuel	[4]
Mill	Mill	Installing a vertical roller mill for solid fuel	[4]
Mill	Mill	Installing an impact mill for solid fuel	[4]
Mill	Mill	Installing a roller press for solid fuel	[27]
Mill	Separator	Installing a high efficiency separator/classifier	[13]

Table A13. Technological energy-saving solutions for gas treatment.

Exhaust Gas Treatment			
Process Machinery	**Solution Object**	**Energy-Saving Technological Solution**	**Reference**
Cooling tower	Cooling tower	Installing a cooling tower	[4]
Selective non-catalytic reduction system (SNCR)	SNCR	Installing an SNCR (difficult for long kilns)	[4]
Selective catalytic reduction system (SCR)	SCR	Installing a low dust SCR	[4]
Selective catalytic reduction system	SCR	Installing a high dust SCR	[4]
Selective catalytic reduction system	SCR	Replacing existing low dust SCR with high efficiency low dust SCR	[4]
Selective catalytic reduction system	SNCR	Replacing existing SNCRs with high efficiency SNCRs	[4]
Activated carbon	Activated carbon	Installing emission reduction systems based on activated carbon	[4]
Wet scrubber	Wet scrubber	Installation of a wet scrubber system	[4]

Table A14. Technological energy-saving solutions for dedusting.

Dedusting				
Process Machinery	**Solution Object**	**Energy-Saving Technological Solution**	**Reference**	**Comments from Sector Experts**
Electrostatic precipitator (ESP)	ESP	Installing an ESP	[4]	
ESP	Vibration system	Replacing the electrostatic precipitator vibration system with a more modern and efficient one	[4]	
ESP	Electrostatic field generator	Replacing ESP's electrostatic field generator with a more advanced and efficient one	[4]	
ESP	Cleaning system	Installing a vibrating precipitator cleaning system	[4]	
ESP and filters	ESP and filters	Installing a hybrid filter system	[4]	
Dust filtration system	Dust filtration system	Installing an electrofilter	Energy audits	
Dust filtration system	Dust filtration system	Installing a dust filtration system with vertical cylindrical bag fabric filters	[4]	
Dust filtration system	Dust filtration system	Installing a dust filtration system with horizontal pocket fabric filters	[4]	Horizontal bag filters are rarely used
Dust filtration system	Cleaning system	Installing a low-pressure pulse jet cleaning system for dust filter systems with fabric filters	[4]	
Separator	Separator	Installing a high-efficiency air separator at the exhaust gas outlet of the cooler	[4]	

Table A15. Technological energy-saving solutions for other transport systems.

Other Transport Systems			
Process Machinery	**Solution Object**	**Energy-Saving Technological Solution**	**Reference**
Transport systems	Transport systems	Installing a cup elevator to feed raw materials into the mill	[27]
Transport systems	Transport systems	Installing a conveyor belt to feed raw materials into the tank	[9]
Transport systems	Transport systems	Installing a cup elevator to feed the blending silo	[15]
Transport systems	Transport systems	Installing a conveyor belt to feed clay into the wash/mill	[9]
Transport systems	Transport systems	Installing efficient transport systems to handle material entering and leaving the crusher	[17]
Transport systems	Transport systems	Installing a conveyor belt to feed the dryer	[9]
Transport systems	Conveying systems	Installing a cup lifter to feed raw materials into the kiln	[15,27]
Transport systems	Conveying systems	Installing a conveyor belt to feed the oven	[9]
Transport systems	Conveying systems	Installing a conveyor belt to bring material to the pelletizer	[9]
Transport systems	Conveying systems	Installing a conveyor belt to feed preheater/calciner	[9]

Table A16. Technological energy-saving solutions for heat recovery systems.

Heat Recovery Systems			
Process Element	**Solution Object**	**Energy-Saving Technological Solution**	**Reference**
Heat recovery systems	Heat recovery systems	Installing a high-efficiency air separator heat exchanger on the exhaust gas outlet from the cooler	[4]
Heat recovery systems	Heat recovery systems	Installing a steam turbine for electricity production	[35]
Heat recovery systems	Heat recovery systems	Installing an ORC turbine for electricity production	[35]
Heat recovery systems	Heat recovery systems	Installing a turbine for electricity production using the Kalina cycle	[35]
Heat recovery systems	Kiln	Installing heat exchanger to recover heat from the flue gases and pre-heat the thermal oil for the kiln fuel	Energy audits
Heat recovery systems	Kiln	Installing heat recovery systems to use the hot gases leaving the kiln to dry raw materials	[4]
Heat recovery systems	Cooler	Installing systems to recover heat from the thermal waste of the cooler (useful for heating offices or other)	[4]

Table A17. Technological energy-saving solutions for auxiliary systems.

Auxiliary Systems			
Process Machinery	**Solution Object**	**Energy-Saving Technological Solution**	**Reference**
Engines	Engines	Installing variable speed motors (motors with inverters)	[15,16,27]
Engines	Engines	Installing efficient electric motors (class IE2, IE3, or IE4)	[15,16,27]
Engines	Engines	Resetting the pre-tensioning of the motor belts	[21]
Engines	Engines	Re-phasing the motors (three-phase) to rebalance the loads of each phase using capacitors	[17]
Engines	Engines	Replacing motor belts with newer, more efficient ones to optimize tensions. For example, replacing V-belts with toothed belts (preferably for high torques)	[17,21]
Engines	Engines	Rewinding motors already in the system	[17]
Engines	Engines	Installing inverters in motors	[17]

Table A17. *Cont.*

Auxiliary Systems			
Process Machinery	**Solution Object**	**Energy-Saving Technological Solution**	**Reference**
Engines	Engines	Replacing belt motors with hydraulic motors or gearboxes	[17]
Engines	Engines	Re-wiring the engines	[9]
Engines	Engines	Installing motors correctly sized in relation to the power required by the system	[17,18]
Pressure systems	Pressure systems	Replacing compressors, air supply and refrigeration systems with more advanced and efficient machinery	[17,34,43]
Pressure systems	Pressure systems	Installing inverters on compressors, pumps, or fans (i.e., kiln fans)	[17,34]
Pressure systems	Pressure systems	Select turbomachinery appropriate to plant requirements	[4,18]
Pressure systems	Pressure systems	Installing tanks and systems for storing excess compressed air	[4,44]
Pressure systems	Compressors	Replacing compressors needed to cool, suck, stir or inflate glass with fans and blowers	[17,34]
Pressure systems	Pressure systems	Insulating pipes, valves, and pumps and installing sealing to reduce air leakage	[4]
Pressure systems	Compressors	Installing gaskets or other devices or replacing damaged components to reduce air leakage at pipe joints and leakage points	[4,17,34,43,45]
Pressure systems	Compressors	Replacing compressors needed to clean or remove debris with brushes, vacuum pumps, or blowers	[17,34]
Pressure systems	Compressors	Replacing compressors needed to move components with electric or hydraulic actuators	[17,34]
Pressure systems	Compressors	Replacing compressors needed to create a vacuum with vacuum pumps	[17,34]
Pressure systems	Compressors	Replacing compressors to power machines, tools, and actuators with electric motors	[17,34]
Pressure systems	Compressors	Installing compressors powered by electricity (with lower maintenance costs, longer service life and less down-time)	[4,18]
Pressure systems	Compressors	Installing gas-fueled compressors (for which it is easier to obtain variable-speed machines and for which lower plant operating costs are obtained)	[4,18]
Pressure systems	Compressed air distribution systems	Correctly size fittings, filters, and hoses to minimize air leaks	[45]
Pressure systems	Compressed air distribution systems	Fitting valves and pressure regulators in compressed air distribution systems to regulate air supply or shut it off when the machinery is not in use	[17,34]
Pressure systems	Compressed air distribution systems	Installing pipes with the largest possible diameter in air distribution systems to reduce losses	[4,18]
Pressure systems	Compressor nozzles	Replacing and renewing compressed air delivery nozzles (which may be worn, clogged, or corroded)	[4,18]
Pressure systems	Ventilation system fans	Replace gas handling fans with more efficient ones	[17,34,43]
Pressure systems	Heat exchangers	Installing heat exchangers or other systems to recover heat from the compressors	[43,45]
Pressure systems	Silo cement extraction plant	Replacing pumps with rotocells	Energy audits
Pressure systems	Chilled water distribution systems	Installing shut-off valves on the cooling water branches (pushed by pumps), to block flows when the system is at a standstill	Energy audits
Electricity transformers	Electricity transformers	Optimizing transformer losses in the electrical cabin	[17,34]
Electricity transformers	Electricity transformers	Renewing transformers in the electrical cabin (preferably installing k-factor transformers)	Energy audits
Electricity transformers	Electricity transformers	Replacing oil transformers with resin transformers (having less leakage)	Energy audits

References

1. Xu, J.-H.; Fleiter, T.; Eichhammer, W.; Fan, Y. Energy consumption and CO2 emissions in China's cement industry: A perspective from LMDI decomposition analysis. *Energy Policy* **2012**, *50*, 821–832. [CrossRef]
2. Atmaca, A.; Yumrutaş, R. Analysis of the parameters affecting energy consumption of a rotary kiln in cement industry. *Appl. Therm. Eng.* **2014**, *66*, 435–444. [CrossRef]
3. Kermeli, K.; Edelenbosch, O.Y.; Crijns-Graus, W.; Van Ruijven, B.J.; Mima, S.; Van Vuuren, D.P.; Worrell, E. The scope for better industry representation in long-term energy models: Modeling the cement industry. *Appl. Energy* **2019**, *240*, 964–985. [CrossRef]
4. Schorcht, F.; Kourti, I.; Scalet, B.M.; Roudier, S.; Delgado Sancho, L. Best Available Techniques (BAT) Reference Document for the Production of Cement, Lime and Magnesium Oxide. In *Industrial Emissions Directive 2010/75/EU*; Joint Research Centre, European Commission: Seville, Spain, 2013; pp. 1–506.
5. Vinci, G.; D'Ascenzo, F.; Esposito, A.; Musarra, M.; Rapa, M.; Rocchi, A. A sustainable innovation in the Italian glass production: LCA and Eco-Care matrix evaluation. *J. Clean. Prod.* **2019**, *223*, 587–595. [CrossRef]
6. Ishak, S.A.; Hashim, H. Low carbon measures for cement plant—A review. *J. Clean. Prod.* **2015**, *103*, 260–274. [CrossRef]
7. Su, T.-L.; Chan, D.Y.-L.; Hung, C.-Y.; Hong, G.-B. The status of energy conservation in Taiwan's cement industry. *Energy Policy* **2013**, *60*, 481–486. [CrossRef]
8. Alghadafi, E.M.; Latif, M. Simulation of a libyan cement factory. In Proceedings of the World Congress on Engineering (WCE), London, UK, 30 June–2 July 2010; Volume 3, pp. 2292–2296.
9. Madlool, N.; Saidur, R.; Rahim, N.; Kamalisarvestani, M. An overview of energy savings measures for cement industries. *Renew. Sustain. Energy Rev.* **2013**, *19*, 18–29. [CrossRef]
10. Moray, S.; Throop, N.; Seryak, J.; Schmidt, C.; Fisher, C.; D'Antonio, M. Energy efficiency opportunities in the stone and asphalt industry. In Proceedings of the Twenty-Eighth Industrial Energy Technology Conference, New Orleans, LA, USA, 9–12 May 2006; pp. 71–83.
11. Alsop, P.A. *Cement Plant Operations Handbook: For Dry Process Plants*; Tradeship Publications Ltd.: Surrey, UK, 2007.
12. Shanmugam, V.; Natarajan, E. Experimental investigation of forced convection and desiccant integrated solar dryer. *Renew. Energy* **2006**, *31*, 1239–1251. [CrossRef]
13. Brunke, J.-C.; Blesl, M. Energy conservation measures for the German cement industry and their ability to com-pensate for rising energy-related production costs. *J. Clean. Prod.* **2014**, *2*, 94–111. [CrossRef]
14. Lynskey, G. Blending/Homogenizing Silos–All They're Cracked up to Be? In Proceedings of the 2019 IEEE-IAS/PCA Cement Industry Conference (IAS/PCA), St. Louis, MO, USA, 28 April–2 May 2019; pp. 1–4. [CrossRef]
15. Hasanbeigi, A.; Menke, C.; Therdyothin, A. The use of conservation supply curves in energy policy and economic analysis: The case study of Thai cement industry. *Energy Policy* **2010**, *38*, 392–405. [CrossRef]
16. Hasanbeigi, A.; Morrow, W.; Masanet, E.; Sathaye, J.; Xu, T. Energy efficiency improvement and CO2 emission reduction opportunities in the cement industry in China. *Energy Policy* **2013**, *57*, 287–297. [CrossRef]
17. Worrell, E.; Galitsky, C.; Price, L. *Energy Efficiency Improvement and Cost Saving Opportunities for Cement Making', Lbnl-54036-Revision*; Ernest Orlando Lawrence Berkeley National Laboratory, University of California: Berkeley, CA, USA, 2008.
18. Institute for Industrial Productivity. Explore Energy Efficiency Technologies across the Industrial Sectors. 2016. Available online: http://www.iipinetwork.org/ (accessed on 11 January 2021).
19. Saha, B.K.; Chakraborty, B. Utilization of low-grade waste heat-to-energy technologies and policy in Indian industrial sector: A review. *Clean Technol. Environ. Policy* **2016**, *19*, 327–347. [CrossRef]
20. Bond, J.; Coursaux, R.; Worthington, R. Blending systems and control technologies for cement raw materials. *IEEE Ind. Appl. Mag.* **2000**, *6*, 49–59. [CrossRef]
21. Fischer, R. Crusher and screen drives for the mining, aggregate and cement industries. In Proceedings of the IEEE Cement Industry Technical Conference, Dallas, TX, USA, 10–14 May 1992; pp. 108–147. [CrossRef]
22. Fujimoto, S. Modern technology impact on power usage in cement plants. *IEEE Trans. Ind. Appl.* **1994**, *30*, 553–560. [CrossRef]
23. Hasanbeigi, A.; Price, L.; Lin, E. Emerging energy-efficiency and CO2 emission-reduction technologies for cement and concrete production: A technical review. *Renew. Sustain. Energy Rev.* **2012**, *16*, 6220–6238. [CrossRef]
24. Mokhtar, A.; Nasooti, M. A decision support tool for cement industry to select energy efficiency measures. *Energy Strat. Rev.* **2020**, *28*, 100458. [CrossRef]
25. Huang, Y.-H.; Chang, Y.-L.; Fleiter, T. A critical analysis of energy efficiency improvement potentials in Taiwan's cement industry. *Energy Policy* **2016**, *96*, 14–26. [CrossRef]
26. Wang, Y.; Li, K.; Gan, S.; Cameron, C. Analysis of energy saving potentials in intelligent manufacturing: A case study of bakery plants. *Energy* **2019**, *172*, 477–486. [CrossRef]
27. Hasanbeigi, A.; Price, L.; Lu, H.; Lan, W. Analysis of energy-efficiency opportunities for the cement industry in Shandong Province, China: A case study of 16 cement plants. *Energy* **2010**, *35*, 3461–3473. [CrossRef]
28. Supino, S.; Malandrino, O.; Testa, M.; Sica, D. Sustainability in the EU cement industry: The Italian and German experiences. *J. Clean. Prod.* **2016**, *112*, 430–442. [CrossRef]
29. Sustainability Report Federbeton. Federbeton Confindustria. 2019. Available online: https://www.federbeton.it/Pubblicazioni (accessed on 3 November 2020).

30. Oggioni, G.; Riccardi, R.; Toninelli, R. Eco-efficiency of the world cement industry: A data envelopment analysis. *Energy Policy* **2011**, *39*, 2842–2854. [CrossRef]
31. Omer, A.M. Energy use and environmental impacts: A general review. *J. Renew. Sustain. Energy* **2009**, *1*, 53101. [CrossRef]
32. Directive 2012/27/EU of the European Parliament and of the Council of 25 October 2012 on energy efficiency, amending Directives 2009/125/EC and 2010/30/EU and repealing Directives 2004/8/EC and 2006/32/EC Text with EEA relevance, Official Journal of the European Union. 2012. Available online: https://eur-lex.europa.eu/legal-content/EN/TXT/?uri=celex%3A32012L0027 (accessed on 29 March 2021).
33. National Agency for New Technologies, Energy, and Sustainable Economic Development (ENEA). 9th Annual Energy Efficiency Report (Rapporto Annuale sull'Efficienza Energetica-RAEE). Available online: https://www.efficienzaenergetica.enea. it/pubblicazioni/raee-rapporto-annuale-sull-efficienza-energetica/rapporto-annuale-sull-efficienza-energetica-2021.html (accessed on 13 December 2020).
34. Galitsky, C.; Worrell, E.; Masanet, E.; Graus, W. *Energy Efficiency Improvement and Cost Saving Opportunities for the Glass Industry. An ENERGY STAR® Guide for Energy and Plant Managers'*; University of California, Berkeley National, Laboratory: Berkeley, CA, USA, 2008. [CrossRef]
35. European Cement Research Academy (ECRA). Development of State of the Art-Techniques in Cement Manufacturing: Trying to Look Ahead. In *CSI/ECRA-Technology Papers 2017*; European Cement, Research Academy GmbH, Dusseldorf, Geneva, March 2017; pp. 1–190. Available online: https://www.wbcsdcement.org/index.html (accessed on 18 January 2021).
36. De Carlo, F.; Schiraldi, M.M. Sustainable choice of the location of a biomass plant: An application in Tuscany. *Int. J. Eng. Technol.* **2013**, *5*, 4261–4272.
37. Cantini, A.; De Carlo, F.; Tucci, M. Application of the lean layout planning system in a leather bags manufacturing plant and proposal of an approach to engage the company's staff in the research of the layout solution. In Proceedings of the Summer School Web, Bergamo, Italy, 9–11 September 2020.
38. International Eneergy Agency. *Driving Energy Efficiency in Heavy Industries*; IEA: Paris, France, 2021; Available online: https://www.iea.org/articles/driving-energy-efficiency-in-heavy-industries (accessed on 29 March 2021).
39. Pozzetti, A. Appunti di tecnologie industriali. In *Rielaborazione Delle Lezioni del Prof Pozzetti*; CUSL (Milano): Milan, Italy, 2003; pp. 1–354, ISBN 8881321823.
40. Cement Equipment Corp. Infinity for Cement Equipment. 2021. Available online: https://www.cementequipment.org/home-page/ (accessed on 11 January 2021).
41. Van Puyvelde, D.R. Modelling the hold up of lifters in rotary dryers. *Chem. Eng. Res. Des.* **2009**, *87*, 226–232. [CrossRef]
42. Duda, W.H. *La Fabbricazione del Cemento*; ET Edizioni Tecniche: Milan, Italy, 1976.
43. Benedetti, M.; Bonfa', F.; Bertini, I.; Introna, V.; Ubertini, S. Explorative study on Compressed Air Systems' energy efficiency in production and use: First steps towards the creation of a benchmarking system for large and energy-intensive industrial firms. *Appl. Energy* **2018**, *227*, 436–448. [CrossRef]
44. Scalt, B.M.; Garcia Munoz, M.; Sissa, A.; Roudier, S.; Delgado Sancho, L. *Best Available Techniques (BAT) Reference Document for the Manufacture of glass: Industrial Emissions Directive 2010/75/EU (Integrated Pollution Prevention and Control)*; Institute for Prospective Technological Studies (Joint Research Centre), European Commission: Seville, Spain, 2012; pp. 1–485. [CrossRef]
45. Blaustein, E.; Radgen, R. *Compressed Air Systems in the European Union*; European Commission Project Report; LOG_X Verlag GmbH: Stuttgart, Germany, 2000; ISBN 3-932298-16-0.

Article

The Mining and Technology Industries as Catalysts for Sustainable Energy Development

Katundu Imasiku [1,*] and Valerie M. Thomas [2,3]

[1] African Center of Excellence in Energy for Sustainable Development, University of Rwanda, Kigali 4285, Rwanda

[2] School of Industrial and Systems Engineering, Georgia Institute of Technology, Atlanta, GA 30322, USA; valerie.thomas@isye.gatech.edu

[3] School of Public Policy, Georgia Institute of Technology, Atlanta, GA 30322, USA

* Correspondence: katunduimasiku@gmail.com

Received: 1 October 2020; Accepted: 24 November 2020; Published: 12 December 2020

Abstract: The potential for mining companies to contribute to sustainable energy development is characterized in terms of opportunities for energy efficiency and support of electricity access in mining-intensive developing countries. Through a case study of the Central African Copperbelt countries of Zambia and the Democratic Republic of Congo, energy efficiency opportunities in copper operations and environmental impact of metal extraction are evaluated qualitatively, characterized, and quantified using principles of industrial ecology, life cycle assessment, and engineering economics. In these countries the mining sector is the greatest consumer of electricity, accounting for about 53.6% in the region. Energy efficiency improvements in the refinery processes is shown to have a factor of two improvement potential. Further, four strategies are identified by which the mining and technology industries can enhance sustainable electricity generation capacity: energy efficiency; use of solar and other renewable resources; share expertise from the mining and technology industries within the region; and take advantage of the abundant cobalt and other raw materials to initiate value-added manufacturing.

Keywords: copper; cobalt; mining; sustainable energy development; engineering economics; multi-national enterprises

---

1. Introduction

Many of the world's poorest countries rely on mining and fossil fuel extraction as the primary basis of their economy [1]. Development pathways for resource-intensive economies are a core challenge of sustainable development [2]. Despite the great wealth of the materials mined and exported, leveraging these industries into economic benefits for the people has been a continuing challenge. Less resource-intensive countries, together with all the world's rich and middle-income countries, have developed through a transition to a manufacturing economy using fossil fuels countries [3]. However, the manufacturing pathway for developing economies might not be realistic for all countries [4]. A key part of this new development model is vested in the role of multi-national enterprises in supporting sustainable economic development in these regions.

The electronics industry depends on minerals from a number of the foremost economically challenged areas of the world. The industry has in recent years taken responsibility for its supply chains in areas characterized by conflict, child labor, and widespread environmental impacts from mining. Some of the world's largest mining firms have developed a framework for supporting sustainable development in its operations and supply chains [5].

To date, the key topics addressed by multinational sustainable development efforts have been child labor, conflict, the rights of indigenous people, human rights, fair trade, and the environmental impacts of mining, extraction and refining.

This study introduces the concept that mining multi-national enterprises (MNEs) can improve their operations and contribute to energy systems for sustainable development. Energy and its accessibility are at the core of social, economic and environmental concerns facing all nations, especially in developing countries. In several countries, mining and ore refining is the largest electricity consumer. The study explores the potential role of multinationals in energy development through two case studies: Zambia and the Democratic Republic of Congo (DRC), which is also referred to as Central African Copperbelt (CAC). This region is one of the world's richest mineral ore body. It is the world's second largest copper cathode producer and the world's largest cobalt producing region [6,7]. In Zambia, copper and cobalt are the major exports, accounting for more than 70% of foreign exchange earnings. This study explores options for increasing energy efficiency, renewable energy deployment, and providing technical support for energy system reliability and development.

A number of previous studies have addressed the relationship between development and environmental impacts in broad statistics analyses. Aldieri et al. found that there is a spillover effect, that is, learning from proximity to technologically advanced firms, that can support technical efficiency, economic viability, and higher productivity [8]. A review across developed and developing countries by Wang et al. finds that the studies on developed countries predominantly take a multi-industry perspective, while studies from developing countries have focused primarily on manufacturing [9].

There have been some studies on the mining industry in developing countries. Lane's study on the need for collaborations to achieve sustainable mining recognized that for the mining sector to remain sustainable, the concept of inclusive growth needs to be adopted by all players. However, the mining sector in Africa comes with challenges like social equity, social license to operate, local supplier development, new investment model development, the requirement for mines to adopt initiatives for infrastructure and energy development [10]. Lebre et al. identify the co-occurrence of environmental, social, and governance risk factors in mining for energy-relevant metals [2].

This research study evaluates the potential for improved energy systems and energy management in mining to support sustainable development. In contrast to previous work, this study quantifies energy efficiency potential for the copper industry in Zambia and DRC, and suggests additional technical measures for increased use of renewables and sharing of technical expertise to support electrification and development. Based in the field of industrial ecology [11], this work begins to fill the literature gap on quantitative assessments of how the multi-national mining and technology industries can contribute to clean energy access in developing countries.

The field of industrial ecology takes the perspective that industry can be an agent of change in meeting environmental objectives, alongside worker safety and customer satisfaction [11]. Taking a largely engineering and technical perspective, the emphasis is on identifying opportunities for meeting broad societal and environmental objectives through innovation and efficiency. Within industrial ecology, life cycle assessment is an important method for evaluating environmental impacts throughout the entire supply chain of a product or service. This is combined with engineering economic or techno-economic assessment, which evaluates the costs and benefits of different technology or policy options.

The study qualitatively characterizes and quantifies the need for energy efficiency in mine production and shows the potential for mining and technology companies to contribute to the sustainable energy development using principles of industrial ecology, life cycle assessment and engineering economics.

Although framed in an industrial ecology context, the full challenge of energy development must be shaped by local choices and decisions by citizens and governments. By developing an industrial ecology framework for the energy development challenge, we aim to shape the potential for multi-nationals to participate as partners with governments, citizens, agencies, and local entrepreneurs in meeting sustainable development goals.

2. Methods

To achieve this aim, a fourfold method is adopted to first evaluate the current energy scenario in Zambia and DRC including modeling energy flow diagrams for the two cases and second, to benchmark with the leading global copper producer of mined copper, Chile to evaluate the energy requirement for copper production in Zambia and DRC. This will help to quantify the copper production efficiencies for the two main copper production methods in use, pyrometallurgical and solvent extraction (SX) and electrowinning (EW). Thirdly, the study evaluates the environmental impact of metal extraction in terms of greenhouse gas (GHG) emissions per ton of copper production while benchmarking with other top copper producers like Australia and the expected global average GHG emission quantity. Fourthly, the sharing of expertise from mining and technology industries for energy development was explored because these tech industry experts can serve on advisory panels for policy formulation and provide a knowledge base for possible replication within Sub-Saharan Africa.

Figure 1 gives a visual perspective of the adopted research method to demonstrate how the mine technology industry can be used to support sustainable energy development while fostering economic growth and development in SSA.

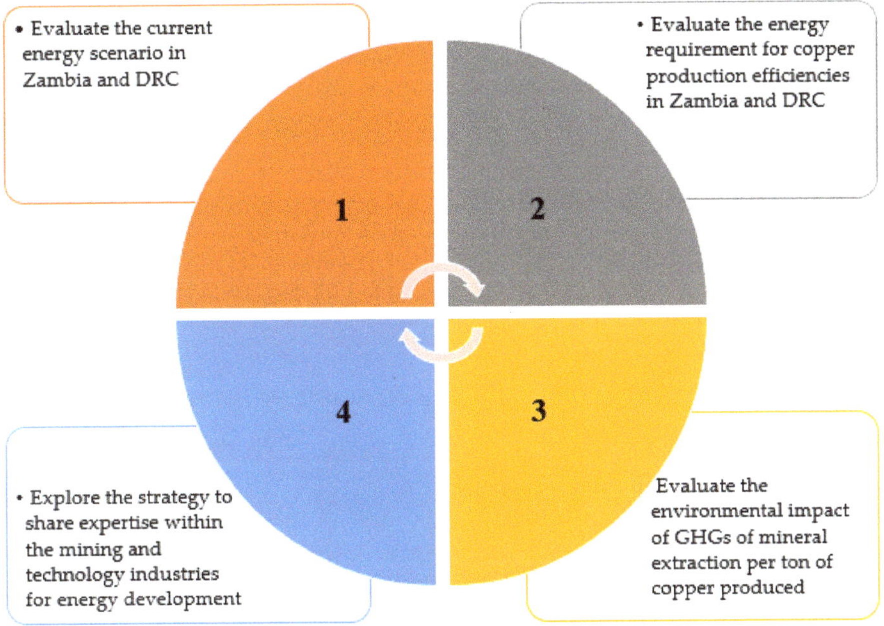

Figure 1. Evaluation steps for mine technology industry support for sustainable energy development.

First, the Central African Copperbelt is a mineral-rich region, but each location has a different political governance system, energy management systems and challenges which may need a different approach and methodology. However, the mining industry in CAC has synergies like having some mine MNEs operating in both Zambia and the DRC as mine owners. This background coined the evaluation of the energy scenarios in DRC and Zambia separately using energy flow diagrams (Sankey) to show the produced energy, its conversion or transformation visually and quantitatively including primary energy, imports, exports and losses and energy consumption by economic sectors in gigawatt-hours (GWh).

Second, the energy requirement for copper production efficiencies in Zambia and DRC was then evaluated to while benchmarking the engineering economic analysis copper production per ton for Zambia and DRC with Chile, the leading global producer of mined copper. The total annual smelting capacities, measured in tons per annum, was also estimated in CAC.

Third, the environmental impact of GHGs of mineral extraction per ton of copper produced was analyzed while benchmarking with another top copper producer, Australia to qualitatively analyze the greenhouse gas emission contribution in the mining and milling sector and smelting sector according to the energy resource used in the different mining activity or process.

Finally, the need to share this expertise for broader energy development projects could make a significant impact on the energy and economic development in DRC and Zambia. Apart from benchmarking with Chile, which has already deployed solar energy on a large scale as a main source of electricity for copper production, this stage explores the strategy to share expertise within the mining and technology industries for energy development.

3. Results

3.1. Energy in Zambia and the Democratic Republic of Congo

Zambia's electricity generation is dominated by hydropower that comprises over 95% of total generation capacity. About 90% of hydro generation comes from just two power stations, the Kariba North Bank and the Kafue Gorge, in the country's Southern province. The country additionally generates about 0.5 GWh using diesel generation system for standby supply. Zambia's sole focus on hydroelectricity is understandable but makes it vulnerable to drought. Zambian power plants are summarized in Table A1 (Appendix A).

Zambia has an average national electrification rate of about 31%, with 61% of the urban areas having access to electricity, while the rural community electrification rate is only 4% [12,13]. Figure 2, of the energy flow for electricity in Zambia, shows that more than 52.3% of the electricity is used by industry, almost all of which is for the metals industry. Zambia's key mine MNEs are Glencore, Mopani, Vedanta Resources, Konkola Copper, First Quantum Minerals, Kalumbila, Barrick Gold and Lumwana mines [14].

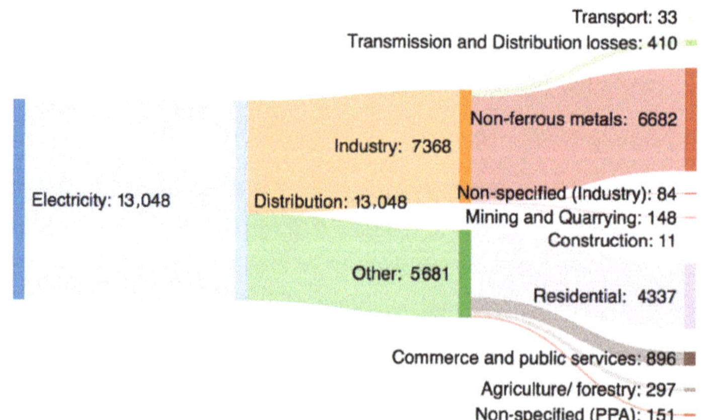

Figure 2. Electricity energy flow in Zambia, as of 2018 (GWh), as built using SankeyMATIC.

The DRC is endowed with large mineral resources and has the potential to install up to 100,000 MW of hydropower capacity. The installed capacity is 2790 MW. Almost 8% of the produced hydropower is consumed by the residential sector while other sectors mostly consume oil. Only 1% of the people in rural areas of DRC and about 8% in the urban areas have access to hydroelectric power [15,16]. The country's electrification rate remains low at 9.6%, and the government's vision is to increase the level of service up to 32% in 2030 [17].

The major mining activities in DRC are in Katanga province whose power capacity is around 900 megawatts (MW) but only about 461.7 MW is available for usage, due to several issues including poor power governance and lack of support from government; lack of findings with high electricity tariffs that are not cost reflective; poor electric utility performance characterized with poor recovery of public consumption electricity invoices that are as high as 40% of total consumption; absence of a regulatory agency and a Rural Electrification Agency; high taxes, and import duties and most importantly the current installed hydropower system has poor infrastructure that leads to many technical losses in the transmission and distribution networks of SNEL. This has resulted in a low distribution efficiency of hydropower with a far lower distributed power compared to the installed power (Figure 3).

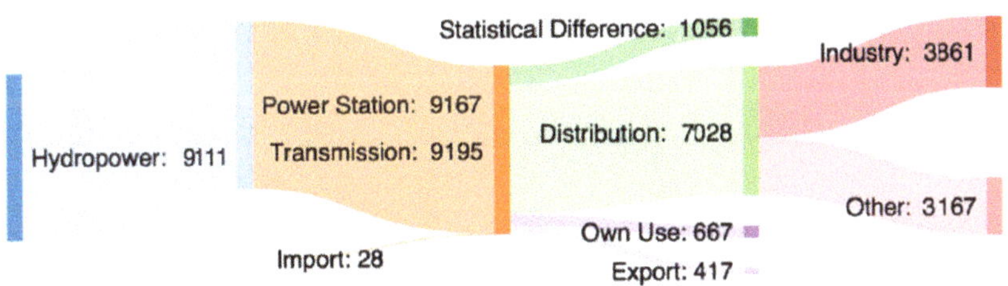

Figure 3. Electricity energy flow in the Democratic Republic of Congo (DRC), as of 2016 (GWh), as built using SankeyMATIC.

The low generation cost of DRC's largest power plant, Inga, potentially reduces the overall cost of electricity in DRC, making it cheap and affordable, but this is subject to other factors like good governing policies and political stability [18,19]. Table A2 (Appendix A) shows the installed, planned electricity capacity under development in DRC as of 2018.

In the past, DRC had huge losses in transmission and distribution; therefore, only 1228 MW (almost 50% of installed capacity) was available capacity in DRC [19]. This situation improved subsequently by 2018 [20]. Figure 3, of the electricity flow in the Democratic Republic of Congo, shows that more than half (54.9%) of the electricity is consumed by the industry sector, which is predominantly mining. The statistical differences in Figure 3 account for transmission losses while rationale of the Other sector comprises 2444.44 GWh for the Residential sector and 722.22 GWh for Commerce and Public Services sector.

3.2. Copper and Cobalt Production Efficiency Potential

Copper production in Zambia and the DRC is benchmarked with Chile, the leading global producer of mined copper. Table 1 shows the energy use for copper production in Chile, Zambia, and the Democratic Republic of Congo. The two main methods used in copper production are pyrometallurgical and solvent extraction (SX) and electrowinning (EW).

Table 1. Energy requirement for copper production in Chile, Zambia and DRC.

	Copper Production (Mt/year)	Electricity for Cu Production (TWh/year)	Production Method: Concentrate (%)	Production Method: SX-EW (%)	Energy Efficiency (kWh/kg Cu)
Chile	5.6	22	63	33	3.9
Zambia	0.755	6.2	57	43	8.2
DRC	0.85	7	90	75	8.2

Energy consumption in copper mining and refining varies with the nature of the ore. Copper production operations in DRC and Zambia has the advantage of high ore grades (3 to 5% acid-soluble Cu), and consequently significantly higher pregnant leach solutions (PLS) copper grades found in other regions [21].

The two main copper extraction processes are the pyrometallurgical process and the solvent extraction and electrowinning (SXEW). The pyrometallurgical processing involves crushing of abundant sulfide ore deposits that are mined and then concentrated into copper concentrate, containing around 30% pure copper which is then smelted into copper anode of 99.5% purity. The copper concentrate utilized in refineries to produce copper blister, which contains about 98% pure copper. The blister is further processed to produce pure copper cathode at copper refineries through leaching, SX and EW to make copper cathodes. The SXEW process is energy intensive (~2.1 MWh/t Cu) because it directly produces copper cathodes from oxide ores. The pyro-metallurgical process is used to recover copper from the mined ore, while SXEW plants process mostly oxide ores but not scrap [22].

The annual smelting capacities are: 870,000 tons per annum at Mufulira smelter, 311,000 tons per annum at Nchanga smelter, 300,000 tons per annum at Nkana Refinery, 150,000 tones of blister copper per annum at Chambishi smelter and 25,000 tons of blister copper at Chambishi Metals refineries and 336,000 tons at Kansanshi smelter [23,24].

The Kansanshi Mine Smelter, identified as the most advanced copper smelter in the world, is worth $900 million and started operating in early 2015 in North-Western province of Zambia. Owned by First Quantum Minerals' (FQM) the Kansanshi Mine Smelter is of strategic importance to both DRC and Zambia. First, the significance is because Zambia's North-Western province has three copper mines which produce more than half of Zambia's copper. Second, since Kansanshi Mines in Zambia and Sentinel copper mine in DRC are both owned by First Quantum Minerals (FQM), the copper concentrate from Sentinel copper mine in nearby Kalumbila in DRC are refined at the new refinery in Zambia where acid is abundant to treat oxide copper ore during the refinery process [25].

More broadly, due to robust domestic smelting capacity, rising mine output and continued support from the Zambian government, the smelting capacity has also improved due to large investment by multi-nationals, making Zambia the largest refined copper producing country in Africa, ahead of DRC and South Africa. The Zambian copper mines are forecast to continue to grow.

The three main competitive refinery technologies are reverberatory furnaces, flash furnaces and heap leach-SX-EW. These three technologies can be compared on the basis of ore grade, scale and other features. Out of these three technologies the heap leach-SX-EW is more modern although it is still being improved. Previously the SX-EW process had an extractant consumption of about 3 kg per ton of copper but this has been improved to about 2 kg per ton of copper. Reverberatory furnaces and flash furnaces can achieve a total recovery of 88%, whereas the SX-EW method achieves a marginal 62%. On the contrary, concerning energy consumption, the flash smelters consume about 6000 kWh/t, and the reverberatory smelters consumes 5700 kWh/t, whereas the SX-EW smelters can consume as much as 9350 kWh/t [26]. The engineering economic analysis to evaluate the energy requirement for copper production is given in Table 1.

Table 1 indicates that copper production in Zambia and DRC could potentially increase its energy efficiency by a factor of two (50%) [27]. A more detailed local analysis would be needed to fully characterize the energy efficiency potential. An interview with a Zambian mineral processing specialist Dr. Kenneth Sichone, on 5 September 2018, confirmed that some MNEs that invested in Zambia in early 2000, after the privatizations of mines, used Chilean processing technology, without considering the concentrate types and copper ore grade against the design specification of the processing technologies and this reduced the production levels [28]. Globally, average production efficiencies can be as low as 60 GJ/t (8.34 kWh/kg) of copper due to low ore grades, ranging to about 30 GJ/t (4.2 kWh/kg) of copper for the most efficient refinery systems. The production efficiency reduces as the ore grade of copper reduces especially to as low as 0.1% [26]. On a global basis, energy contributes approximately 35% of total copper refinery production costs [29].

3.3. Environmental Impact of Metal Extraction

In the Australian copper industry, the main contributor to greenhouse gas emissions are the mining and milling, accounting for about 57% while smelting accounts for about 27% [30]. The extraction of metal ores involving both underground and open-pit mining techniques have different energy consumptions and GHG emissions or waste. The selection of an appropriate processing technique depends on many factors but not restricted to the character of the ore and the location of the ore, its size, at what depth and most importantly the concentrate or ore grade. Unlike the open-pit mining method, underground mining consumes a lot of energy owing to larger demands from hauling, ventilation and water pumping. In Australia, most of the copper ore comes from underground mines [31]. The global average of greenhouse gas (GHG) emissions was estimated in 2012 to be 6.58 t CO_2 per ton of copper production [32].

The CO_2 emissions of copper processing at different refinery stages depend on the energy source used in that stage which could be coal, diesel, natural gas and solar, hydropower or any other renewables energy resource. As ore grade decreases, higher energy consumption is expected for future copper production [32].

Further, to enhance competitiveness by resolving societal problems firms should be credited at country and regional level especially if benchmarked against the international mining policies that govern the regional mining policies [33].

3.4. Expertise from Mining and Technology Industries for Energy Development

In Zambia and the Democratic Republic of Congo (DRC), hydropower is the main source of electricity. As discussed above this contributes to the production of copper and cobalt production with relatively low greenhouse gas emissions.

In Chile, solar power is becoming the main source of electricity for copper production [34]. The Chilean mining region has outstanding solar resources and few other energy resources, making the location particularly well suited for solar power development. However, given the success of Chilean use of solar power in the metals industry, there may be potential to introduce utility-scale solar to the Zambia and DRC mining districts.

The extensive energy consumption of mine operations can be utilized to manage the introduction of intermittent renewables such as wind and solar into the national electricity system [35]. Similar to the use of electric vehicle charging to buffer the variability of solar power [36], the operations of mining can also provide flexibility to reduce their costs and electricity system costs.

Metals and technology industries have developed substantial expertise in electricity systems operation. Sharing this expertise for broader energy development projects could make a significant impact on the energy and economic development in these countries [37].

4. Discussion

The technology industries use copper and cobalt for production of electronics and batteries; these are key contributors to a global sustainable energy future. The production of copper and cobalt in the Democratic Republic of Congo (DRC) and in Zambia does have some strong sustainability features, including use of renewable energy sources in the production process, and increased energy efficiency over time. Even so, at a national level the energy systems of DRC and Zambia are underdeveloped, and electricity access is low. The metals industries are the main users of electricity, and energy costs are significant for them. The situation presents an opportunity for the metals industries and the technology industries to catalyze improved energy access because energy efficient production can save about 50% of energy. This can allow both Zambia and DRC to increase the consumption of other uses to about 75% as opposed to the initial 46.4% average power to sectors other than the mines in CAC.

To design a sustainable energy supply system in CAC within the mines, both generation-mix and energy efficiency are important components. Sustainable mining can make the business more viable because renewables can be cheaper than conventional power. Moreover, improved socio-economic development of both society and the mining firm benefits the nation. Requirements from the customers of mining companies, throughout the international supply chain, are increasingly pushing suppliers to meet sustainability standards. Achieving socio-economic development implies that renewable energy will not simply be used to solve societal problems concerning the environment but support the business model of the international mining industry and potentially provide benefit more reliable energy supply at a reduced cost. From a business perspective, this provides corporate shared value amongst all stakeholders.

The extraction of metal ores involving both underground and open-pit mining techniques have different energy consumptions and GHG emissions or waste. Further, the selection of an appropriate processing technique depends on many factors but not restricted to the character of the ore and the location of the ore, its size, at what depth, and most importantly the concentrate or ore grade. Unlike the open-pit mining method, underground mining consumes a lot of energy owing to larger demands from hauling, ventilation, and water pumping.

The CO_2 emissions of copper processing at different refinery stages depend on the energy source used in that stage which could be coal, diesel, natural gas and solar, hydropower, or any other renewables energy resource. A quantitative analysis of greenhouse gas emissions from cobalt and copper production in Zambia and the DRC is beyond the scope of this study. However, the dominance of hydropower in the electricity systems of both countries suggests that the greenhouse gas emissions would be lower than that produced in Australia, which has a fossil fuel-dominated electricity system. As ore grade decreases, higher energy consumption is expected for future copper production.

The metals and technology industries have developed substantial expertise in electricity systems operation. Chile uses solar power in the metals industry, and this shows that there may be potential to introduce utility-scale solar to the Zambia and DRC mining districts. The combination of solar power with hydropower can increase the overall resilience of the electricity system, mitigating the tendency of hydropower to be affected by drought and seasonal variations. Implementation of state-of-the-technology solar systems for metals development in Zambia and DRC could benefit the metals industries and could also provide a technological foundation for the broader implementation of solar power in these countries and the broader region. Sharing this expertise for broader energy development projects could make a significant impact on the energy and economic development in these countries. Further, the energy ministries and utility systems of these countries could consider inviting the metals industry and tech industry experts to serve on advisory panels and to find ways to support the beneficial development of energy access. Situating the sharing of expertise in the concept of United Nations or other international development agency activities can provide a framework for sharing of information and access to innovative expertise in a way that is transparent and that could be replicated throughout Sub-Saharan Africa and elsewhere.

Other innovative spring-out opportunities that would benefit the local people in DRC and Zambia concern the use of cobalt in lithium-ion batteries and renewable energy technologies, power grid stabilization, and electric vehicles. The mine MNEs in Zambia and DRC export unprocessed copper and cobalt and purchase back finished battery components in CAC to support the energy storage industry. Exploring value-added manufacturing of battery components using readily available raw materials (cobalt and copper) would support sustainable development in CAC.

Multi-national companies in the metals and technology sectors have strong sustainability and social responsibility programs that support the communities and environments in which they operate. Some multi-national enterprises have set high employment standards in compliance with industry best practice and social responsibilities as stipulated by the International Council on Mining and Minerals [7]. Some multi-national firms that source materials from the CAC, notably including Apple, have worked towards reduced environmental impacts and seek to be trendsetters for other multi-nationals [38]. In 2018, Transparency International (TI) worked to ensure that multi-national companies, such as Apple, meet the appropriate labor standards concerning the procurement of cobalt from firms that observe sustainable mining practices and not from child miners in DRC [39,40]. In the energy area, in 2018 Apple declared its complete migration to 100% renewable energy and encouraged other multi-nationals to emulate it [41].

Including a focus on energy development in the countries that supply essential materials, as part of these social responsibility programs, could help transform the energy policies and energy systems of these mining-dependent nations and provide an example of the potential for extractive industries to make positive substantial contributions to development.

5. Conclusions

Mining activities absorb much of the electricity generated in Zambia and the DRC. Copper production efficiency in the CAC is relatively low. With the high ore grades averaging in CAC, the energy efficiency in the refinery processes can be improved by about a factor of two. As this electricity-generating capacity already exists, reducing the energy demand would improve electricity availability in Zambia and DRC at a low cost. The implementation of more energy efficient production would benefit the mining industries by reducing production costs, and would benefit the region as a whole. The key finding, that energy efficiency measures in one industry could substantially improve energy access in two nations, suggests a new avenue to meeting sustainable energy development goals. In Zambia and the DRC, and potentially in other countries, a major portion of future energy needs can be met through energy efficiency rather than entirely by constructing new generating sources.

The mining and technology industries can enhance sustainable electricity generation capacity by deploying energy efficiency strategies throughout the mine operations. They can identify the areas or mine operation activities for renewable resource usage especially the abundant solar energy resource. The intermittency of solar energy could potentially be mitigated with battery energy storage Use of cobalt-containing storage batteries would provide Zambia and DRC with the technological benefits of their mining industries. The mining MNEs can enhance the energy storage systems by taking advantage of the abundant Cobalt raw material to explore battery component manufacturing to boost economic diversity beyond Copper and Cobalt mining in the CAC region. Further, the sharing of expertise from the mining and technology industries within the region can also support sustainable energy development and mining.

The CO_2 emissions of copper processing at different refinery stages depend on the energy source used in that stage which could be coal, diesel, natural gas and solar, hydropower, or any other renewables energy resource. Renewable energy resource generation-mix strategy is encouraged to achieve, net-zero emission energy generation systems within the copper mines in CAC.

This study has used a case study approach to examine the potential for technology and mining companies to contribute to energy development while meeting their own environmental goals. Expanding on this perspective, a comprehensive analysis of the potential for energy efficiency and renewable energy in mining to support both developing county energy goals and technology firms' sustainability goals could provide a basis for concerted programs in clean energy access. In addition, a more detailed quantitative industrial ecology study in the CAC region could quantify energy flows through mine industrial systems and quantify the greenhouse gas emission contributions within the mine operation processes for strategic energy modeling and optimization using a life cycle assessment. This would support future mine operation designs to increase renewable energy applications and exploit the abundant renewable resources as a leapfrog strategy for the MNEs to achieve sustainable energy development.

Author Contributions: Data curation: K.I. and V.M.T.; writing-original draft preparation: K.I.; writing—review and editing: K.I. and V.M.T.; visualization: V.M.T.; supervision formal analysis: V.M.T.; funding acquisition: K.I.; investigation: V.M.T. and K.I.; methodology: K.I.; project administration: V.M.T.; resources: K.I. All authors have read and agreed to the published version of the manuscript.

Funding: This research received no external funding.

Sustainability **2020**, *12*, 10410

Acknowledgments: The authors would like to thank the African Center of Excellence in Energy for Sustainable Development and the University of Rwanda for their support.

Conflicts of Interest: The authors declare no conflict of interest.

Appendix A

Table A1. Installed Generation Capacity in Zambia as of 2018 (Energy Regulation Board of Zambia, 2018).

Power Station	Installed Capacity (GWh)	Generation Type	Operator
Kafue Gorge	5142.56	Hydro	ZESCO
Kariba North Bank	3740.04	Hydro	ZESCO
Kariba North Bank Extension	1870.02	Hydro	ZESCO
Victoria Falls	561.01	Hydro	ZESCO
Lunsemfwa and Mulungushi	290.89	Hydro	LHPC
Maamba	1558.35	Coal	Maamba Collieries
Ndola	571.40	Heavy Fuels	Ndola Energy Plant
IPP Small Hydro **	627.24	Hydro	IPPs
Small Hydro *	225.96	Hydro	ZESCO
Isolated Generation ***	18.7	Diesel	ZESCO
CEC (Stand By)	415.56	Diesel	CEC
(Solar Samfya) (Sinda, Kitwe Samfya)	5.66	Solar	CEC, Muhanya and Rural Electrification Authority
Total Installed Capacity	15,027.39		

Note: * Lusiwasi, Musonda falls, Shiwang'andu, Chishimba falls and Lunzua; ** Zengamina Power Limited and Itezhi-Tezhi; *** Shangombo and Luangwa.

Table A2. Installed generation capacity in DRC by 2018.

Power Station	Capacity (MW)	Generation Type	Status	Name of River	Source
Inga I	351	Run of river	Operational	Congo	[42]
Inga II	1424	Run of river	Operational	Congo	[42]
Nseke	260	Reservoir	Operational	Lualaba	[19,43]
Ruzizi I	40	Reservoir	Operational	Ruzizi	[19]
Ruzizi II	45	Reservoir	Operational	Ruzizi	[19]
Ruzizi III	147	Run of river	Under Construction	Ruzizi	[19]
Ruzizi IV	200	Run of river	Proposed	Ruzizi	[44]
Rutshuru	13.8	Run of river	Operational	Rutshuru	[45]
Mutwanga	10	Run of river	Operational	Semliki	[46]
Nzilo	108	Run of river	Operational	Lualaba	[19,47]
Mobayi	11.5	Reservoir	Operational	Ubangi	[48]
Koni	36	Reservoir	Operational	Lufira	[47]
Mwadingusha	71	Reservoir	Operational	Lufira	[49]
Kyimbi	18	Reservoir/Waterfall	Operational	Kyimbi	[44]
Tshopo	19.65	Reservoir/Waterfall	Operational	Tshopo	[19,48,50]
Kilubi	9.9		Operational		[44]
Lungundi	1.6		Operational	Kasai	[44]
Mpozo	2.21		Operational	Mpozo	[44]
Sanga	12	Reservoir/Waterfall	Operational	Inkisi	[51]
Zongo	75	Reservoir	Operational	Insiki	[19,47]
Zongo II	150	Reservoir	Operational	Insiki	[52]
Katende	64	Run of river	Under Construction	Daugava	[53]
Kakobola	10.5	Run of river	Under Construction	Rufuku	[54]
Grand Inga	39,000	Run of river	Proposed	Congo	[42]
Inga III	4800	Run of river	Under Construction	Congo	[42]
IPPs owned by 14 Mine MNEs	100.3	Hydro-electric	Operational	Various Mine sites	[44]
Other 25 Thermal Plant IPPs	31.6	Thermal Power Stations	Operational	Various sites	[44]
Planned Capacity	39,000.00				
Under Construction	5021.50				
Total Installed	2790.56				

References

1. Venables, A.J. Using Natural Resources for Development: Why Has It Proven So Difficult? *J. Econ. Pespectives* **2016**, *30*, 161. [CrossRef]
2. Lèbre, É.; Stringer, M.; Svobodova, K.; Owen, J.R.; Kemp, D.; Côte, C.; Arratia-Solar, A.; Valenta, R.K. The social and environmental complexities of extracting energy transition metals. *Nat. Commun.* **2020**, *11*, 4823. [CrossRef]
3. International Copper Study Group. *The World Copper Factbook*; ICSG: Lisbon, Portugal, 2018.
4. IMF. *Botswana: Mining a New Growth Model*; IMF: Washington, DC, USA, 2018.
5. Buxton, A. MMSD+10: Re ecting on a Decade of Mining and Sustainable Development. 2012. Available online: https://pubs.iied.org/pdfs/16041IIED.pdf (accessed on 15 August 2020).
6. Kaniki, T.; Tumba, K. Management of mineral processing tailings and metallurgical slags of the Congolese copperbelt: Environmental stakes and perspectives. *J. Clean. Prod.* **2019**, *210*, 1406. [CrossRef]
7. Imasiku, K.; Thomas, V.M.; Ntagwirumugara, E. Unpacking ecological stress from economic activities for sustainability and resource optimization in sub-Saharan Africa. *Sustainability* **2020**, *12*, 3538. [CrossRef]
8. Aldieri, L.; Makkonen, T.; Vinci, C.P. Environmental knowledge spillovers and productivity: A patent analysis for large international firms in the energy, water and land resources fields. *Resour. Policy* **2020**, *69*, 101877. [CrossRef]
9. Wang, C.; Ghadimi, P.; Lim, M.; Ming-Lang, T. A literature review of sustainable consumption and production: A comparative analysis in developed and developing economies. *J. Clean. Prod.* **2019**, *206*, 741–754. [CrossRef]
10. Lane, A. Deloitte: Sustainable Mining in Africa Requires Collaboration. 21 February 2016. Available online: https://www.miningreview.com/top-stories/deloitte-sustainable-mining-in-africa-requires-collaboration/ (accessed on 29 November 2020).
11. Socolow, R.; Andrews, C.; Berkhout, F.; Thomas, V. *Industrial Ecology and Global Change*; Cambridge University Press: Cambridge, UK, 2010.
12. Zambia Development Agency. *Energy Sector Profile*. 2014. Available online: https://ab-network.jp/wp-content/uploads/2014/07/Energy-Sector-Profile.pdf (accessed on 24 November 2020).
13. Energy Regulation Board of Zambia. *Statistical Bulletin*; ERB: Lusaka, Zambia, 2018.
14. Energy World. Zambia Aims for Cost-Reflective Electricity Prices by End 2018. 21 June 2018. Available online: https://energy.economictimes.indiatimes.com/news/power/zambia-aims-for-cost-reflective-electricity-prices-by-end-2018/64681648 (accessed on 13 July 2019).
15. USAID. Democratic Republic of Congo—Power Africa Fact Sheet. 16 April 2020. Available online: https://www.usaid.gov/powerafrica/democratic-republic-congo (accessed on 29 November 2020).
16. International Rivers. Congo's Energy Divide. 9 June 2013. Available online: https://www.internationalrivers.org/resources/congo\T1\textquoterights-energy-divide-factsheet-3413 (accessed on 30 June 2019).
17. Ministry of Planning National Investment Promotion Agency-DRC. *Energy*. 18 September 2018. Available online: https://www.investindrc.cd/en/Energy (accessed on 29 November 2020).
18. Oyewo, A.S.; Farfan, J.; Peltoniemi, P.; Breyer, C. Repercussion of large scale hydro dam deployment: The case of Congo grand inga hydro project. *Energies* **2018**, *11*, 972. [CrossRef]
19. Kadiayi, A. Overview of the Electricity Sector in the Democratic Republic of Congo. 2013. Available online: https://www.usea.org/sites/default/files/event-/Democratic%20Republic%20of%20Congo%20Power%20Sector.pdf (accessed on 13 July 2019).
20. African Energy Commission. Africa Energy Data Base. 2018. Available online: https://afrec-energy.org/Docs/En/PDF/2018/statistics_2018_afrec.pdf (accessed on 17 July 2019).
21. Sole, K.C.; Tinkler, O.S. Copper solvent extraction: Status, operating practices, and challenges in the African Copperbelt. *J. South. Afr. Inst. Min. Metall.* **2016**, *116*, 553. [CrossRef]
22. Lanz, B.; Rutherford, T.F.; Tilton, J.E. Subglobal climate agreements and energy-intensive activities: An evaluation of carbon leakage in the copper industry. *World Econ.* **2013**, *36*, 254–279. [CrossRef]
23. USGS. *Zambia: Estimated Production of Production of Mineral Commodities*; USGS Mineral Information Summary: Reston, VA, USA, 2014.
24. ZCCM Investment Holdings Plc. Substantial Cobalt and Copper Reserves and Mining, Refining and Tolling of Cobalt and Copper. 2018. Available online: http://www.zccm-ih.com.zm/copper-cobalt-gold/chambishi-metals/ (accessed on 24 November 2020).

25. KPMG Global Mining Institute. Democratic Republic of Congo Country Mining Guide. 2014. Available online: https://assets.kpmg/content/dam/kpmg/pdf/2014/09/democratic-republic-congo-mining-guide.pdf (accessed on 29 November 2020).

26. Ayres, R.U.; Ayres, L.W.; Råde, I. *The Life Cycle of Copper, its Co-Products and By-Products*; Springer: Amsterdam, The Netherlands, 2003; pp. 13–21.

27. Fagerström, C. Copper Mining in Chile and its Electric Power Demand. Bachelor's Thesis, Mechanical and Production Engineering, Novia University of Applied Sciences, Vaasa, Finland, 2015. Available online: https://www.theseus.fi/bitstream/handle/10024/91170/Fagerstrom_Christoffer.pdf;sequence=1 (accessed on 24 November 2020).

28. Sichone, K. Interviewee, Senior Lecturer Mines Department, University of Rwanda. 5 September 2018.

29. Cantallopts, J. *Copper Concentrates: Smelting Technologies Update and Cucons Market*; CochilcoChile: Santiago, Chile, 2017; Available online: https://www.cochilco.cl/Presentaciones%20Ingls/Copper%20Concentrates%202017.pdf (accessed on 24 November 2020).

30. Mudd, G.M.; Weng, Z.; Memary, R.; Northey, S.A.; Giurco, D.; Mohr, S.; Mason, L. *Future Greenhouse Gas Emissions from Copper Mining: Assessing Clean Energy Scenarios*; Institute for Sustainable Futures, University of Technology, Civil Engineering: Sydney, Australia, 2012.

31. Haque, N.; Norgate, T. Energy and greenhouse gas impacts of mining and mineral processing operations. *J. Clean. Prod.* **2009**, *18*, 267.

32. McLellan, B.; Corder, G.; Giurco, D.; Ishihara, K. Renewable energy in the minerals industry: A review of global potential. *J. Clean. Prod.* **2012**, *32*, 36. [CrossRef]

33. Hamann, R.; Kapelus, P. Corporate social responsibility in mining in Southern Africa: Fair accountability or just greenwash? *Development* **2004**, *47*, 85. [CrossRef]

34. Vyhmeister, E.; Muñoz, A.C.; Miquel, J.M.; Moya, J.P.; Guerra, C.F.; Mayor, L.R.; Reyes-Bozo, L. A combined photovoltaic and novel renewable energy system: An optimized techno-economic analysis for mining industry applications. *J. Clean. Prod.* **2017**, *149*, 999. [CrossRef]

35. Machalek, D.; Landen, A.Y.; Blackburn Rogers, P.; Powell, K.M. Mine operations as a smart grid resource: Leveraging excess process storage capacity to better enable renewable energy sources. *Miner. Eng.* **2020**, *145*, 106103. [CrossRef]

36. Mouli, C.G.; Bauer, P.; Zeman, M. System design for a solar powered electric vehicle charging station for workplaces. *Appl. Energy* **2016**, *168*, 434. [CrossRef]

37. Imasiku, K.; Valerie, T.; Etienne, N. Unravelling green information technology systems as a global greenhouse gas emission game-changer. *Adm. Sci.* **2019**, *9*, 43. [CrossRef]

38. Apple. Environmental Responsibility Report. 2019. Available online: https://www.apple.com/environment/pdf/Apple_Environmental_Responsibility_Report_2019.pdf (accessed on 29 July 2019).

39. Cuthbertson, A. Apple Faces Child Labor Scrutiny as It Looks to Take Charge of Cobalt Mines. Available online: https://www.newsweek.com/apple-faces-child-labor-scrutiny-it-looks-take-charge-cobalt-mines-815981 (accessed on 6 July 2019).

40. Kambani, S.M. Small-scale mining and cleaner production issues in Zambia. *J. Clean. Prod.* **2003**, *11*, 141. [CrossRef]

41. Sullivan, M. Apple Now Runs on 100% Green Energy, and Here's How It Got There. 2018. Available online: https://www.fastcompany.com/40554151/how-apple-got-to-100-renewable-energy-the-right-way (accessed on 7 June 2019).

42. International Hydropower Association. Democratic Republic of the Congo. August 2015. Available online: https://www.hydropower.org/country-profiles/democratic-republic-of-the-congo (accessed on 30 November 2020).

43. GE—Renewable Energy. Reliable Energy to Kick-Start the Economy. 2019. Available online: https://www.ge.com/renewableenergy/stories/nseke-reliable-energy (accessed on 1 December 2019).

44. Electricity in Africa. Electricity Alert—Electricity in Africa and the Future. 2019. Available online: http://electricityalert.blogspot.com/p/democratic-republic-of-congo.html (accessed on 30 November 2020).

45. World Wide Fund for Nature—WWF. Construction Begins on New Virunga Hydropower Plant. 19 December 2013. Available online: http://wwf.panda.org/?213450/Construction-begins-on-new-Virunga-hydropower-plant (accessed on 30 November 2020).

46. Water Power and Dam Construction. India Helps Fund Congo Hydro Project. 2014. Available online: https://www.waterpowermagazine.com/news/newsindia-helps-fund-congo-hydro-project-4311742 (accessed on 1 December 2020).

47. Global Energy Obervertory. 2011. Available online: http://globalenergyobservatory.org/geoid/42626 (accessed on 29 November 2020).

48. African Energy. *African News Letter*. 27 July 2017. Available online: https://www.africa-energy.com/article/dr-congo-mobayi-mbongo-rehabilitation (accessed on 1 December 2020).

49. Poindexter, G. Mwadingusha Hydro Plant Rehab Supports Developing World's Largest Copper Find in Africa. 2018. Available online: https://www.hydroreview.com/2018/01/09/mwadingusha-hydro-plant-rehab-supports-developing-world-s-largest-copper-find-in-africa/#gref (accessed on 1 December 2020).

50. International Commettee for the Red-Cross. Democratic Republic of the Congo: ICRC Completes Major Engineering Project in Kisangani. 30 July 2002. Available online: https://www.icrc.org/en/doc/resources/documents/feature/2002/5cjj7l.htm (accessed on 1 December 2020).

51. Alternative Energy Africa News. Sanga Hydropower Station Connected to Grid. 2019. Available online: https://www.ae-africa.com/read_article.php?NID=8883 (accessed on 1 December 2020).

52. Takouleu, J.M. DRC: Sinohydro Has Finally Completed the Zongo 2 Hydroelectric Dam. 3 July 2018. Available online: https://www.afrik21.africa/en/drc-sinohydro-has-finally-completed-the-zongo-2-hydroelectric-dam/ (accessed on 1 December 2020).

53. REUTERS. DR Congo Signs $280 Mln Hydro Plant Deal with India. 12 July 2011. Available online: https://www.reuters.com/article/ozatp-congo-democratic-hydroelectric-idAFJOE76B0D420110712 (accessed on 1 December 2020).

54. The Infrastracture Consortium for Africa. DRC: An Investment of 42 Million USD for Kakobola Hydroelectric Power Station. 13 January 2011. Available online: https://www.icafrica.org/en/news-events/infrastructure-news/article/drc-an-investment-of-42-million-usd-for-kakobola-hydroelectric-power-station-1533/ (accessed on 1 December 2020).

 *sustainability*

Article

From Program to Practice: Translating Energy Management in a Manufacturing Firm

Mette Talseth Solnørdal * and Elin Anita Nilsen

School of Business and Economics, UiT The Arctic University of Norway, Pb 6050 Langnes,
9037 Tromsø, Norway; elin.nilsen@uit.no
* Correspondence: mette.solnordal@uit.no; Tel.: +47-7764-6016

Received: 28 October 2020; Accepted: 2 December 2020; Published: 3 December 2020

Abstract: A promising way to stimulate industrial energy efficiency is via energy management (EnM) practices. There is, however, limited knowledge on the implementation process of EnM in manufacturing firms. Aiming to fill this research gap, this study explores the implementation of a corporate environmental program in an incumbent firm and the ensuing emergence of EnM practices. Translation theory and the 'travel of management ideas' is used as a theoretical lens in this case study when analysing the process over a period of 10 years. Furthermore, based on a review and synthesis of prior studies, a 'best EnM practice' is developed and used as a baseline when assessing the EnM practices of the case firm. Building on this premise, we highlight four main findings: the pattern of translation dynamics, the key role of the energy manager during the implementation process, the abstraction level of the environmental program and, 'translation competence' as a new EnM practice. Managerial and policy implications, as well as avenues for further research, are provided based on these results.

Keywords: energy efficiency; energy management practices; translating management ideas; case study

1. Introduction

Increasing environmental degradation and risks from disasters have placed the mitigation of climate change and the emission of greenhouse gases (GHG) among the most pressing issues of the twenty-first century. The industrial sector accounts for a large proportion (37%) of the world's total energy consumption [1]. Therefore, increased industrial energy efficiency (EE) is an important means for sustainable development [2], and is essential to reach global sustainability targets such as the Paris Agreement [3] and the European 2030 climate and energy framework [4]. Manufacturing firms can increase their EE by implementing new technological measures in their production processes [5] that require less energy to perform the same functions [6], and by behavioural changes [7]. EE reduces energy costs [8] and increases productivity [9,10] and is positively related to firms' financial performance [11,12] and competitiveness [13]. Nonetheless, research has identified a wide range of barriers for industrial EE [14–16]. Hence, the manufacturing sector's full potential remains unexploited [17–19], leading to an 'EE gap' [20] which denotes the discrepancy between the theoretically optimal and current level of EE.

While the EE gap has mainly been addressed with relevance to technological innovations [21,22], it also consists of behavioural and managerial components, conceptualised as the 'extended EE gap' [7]. Thus, manufacturing firms' enhanced EE improvements and sustainable development [23] require that they accompany technological innovations with the implementation of energy management (EnM) [24–27]. EnM is thus a significant driver for increased EE in manufacturing firms [28,29] and requires academic attention.

Industrial EnM has been thematised in a number of publications, and comprehensively reviewed by e.g., Schulze et al. [30] and May et al. [26]. These reviews illustrate the broad variety in the conceptual understanding of EnM in academic research. Lawrence et al. [25] describe EnM as the procedures in industrial firms addressing energy use to improve EE. Moreover, EnM is described as technical energy monitoring and measurement systems [31], and organizational systems for the continual improvement of energy performance [32]. Furthermore, EnM is considered a tool in helping firms overcome barriers to improving industrial EE [33,34]. Agencies such as the International Organisation for Standardisation (ISO) denote EnM as a standard, such as the ISO 50001 [35] energy management standards, whereas Sannö el al. [27] and Sa et al. [36] describe EnM as a program. EnM can also be incorporated in environmental policy programs at both national and regional levels [22,37]. Despite considering EnM from different perspectives, these studies present EnM as ideal recipes, at various abstraction levels, that firms should commit to for increasing their EE. Furthermore, in a comprehensive definition, EnM Schulze et al. [30] assert that 'EnM comprises the systematic activities, procedures and routines within an industrial company including the elements strategy/planning, implementation/operation, controlling, organisation and culture and involving both production and support processes, which aim to continuously reduce the company's energy consumption and its related energy costs'. This definition endorses the need for transforming EnM into EnM practices in the organisation, at the strategic, operational and human level.

It is worth mentioning, however, is that the contribution of EnM depends on firm-specific and contextual characteristics such as size, energy intensity and production type [7,27], and that these EnM recipes need to be given content and meaning according to the contextual setting of each firm. To this end, there is a broad variety in types of EnM practices [38] that firms can consider. Nonetheless, research shows that manufacturing firms often fail to implement EnM practices and that their EnM maturity level is generally low [8,25,32,39]. This suggests that firms face large challenges when implementing EnM.

In this emerging research field on EnM practices scholars have mainly focused on how to characterise effective and successful EnM practices [16,30,32,33,38,39]. Although these studies provide valuable information for describing and assessing the level of EnM practices in firms [27,36], they do not provide knowledge on the implementation process where EnM programs are transformed into specific EnM practices. This gap in the literature is addressed by Lawrence et al. [25] asserting that 'while barriers to and drivers for industrial EE have been investigated for many industries, there is a lack of studies of barriers to and drivers for EnM practices'. Hence, there is little knowledge on how firms adopt EnM [30] and align EnM practices with firms' core business and strategic agendas [40]. Models to support industrial managers in successfully implementing EnM are also lacking [38]. Furthermore, with the exception of Sannö et al. [27], few empirical studies analyse the implementation of in-house EnM programs in multinational companies (MNC). Hence, there are several calls for more knowledge on the implementation of EnM programs in MNC.

A research tradition that pays attention to the transformation from programs to practices is the 'translation perspective' [41], located within the framework of Scandinavian institutionalism [42–44]. Here, the inherent premise is that implementation is closely associated with the translation of management ideas and models [41,44]. A main focus in this approach has been on variations in how new versions of organisational ideas are translated in the local context. Empirical research has identified that translation takes place in accordance with translation or editing rules, and that 'good translations' foster successful results [45–47]. As such, the outcome of a successful implementation process can be observed in the materialisation of, for instance, new routines and practices. According to this approach one would assume that EnM goes through a transformation from a vague idea to concrete practices as part of implementation. Furthermore, scholars assert that how this transformation is approached with regard to translation will affect if and how the EnM program materialises as definite EnM practices in the organisation.

Using this framework, this study aims to meet these calls by considering the implementation of EnM as a process of translation. We ask: *What is the relationship between the translation process of a corporate environmental program and the successful materialisation of EnM practices in a manufacturing firm?*

The case study is new in the sense that it explores retrospectively over a period of 10 years (2004–2014) the implementation of a corporate environmental program, here called 'EcoFuture'. The data are qualitatively collected in 'Pharma', a subsidiary of an MNC. Through EcoFuture, the global management of the MNC pledged to integrate long-term sustainability objectives into its global business strategy and the business strategies of all subsidiaries.

As such, the study contributes to the EnM literature with new knowledge on the relevance of contextual factors and the role of key translators in reinterpreting and giving the environmental program content in the firm setting. In addition, the results add to our limited knowledge of corporate EnM and EnM practices at the firm level. Moreover, the results indicate the potential of the translation framework in research on EnM, and suggests its use as a 'tool' to gain translation competence to facilitate successful implementation of EnM. This will benefit managers by highlighting specific competences and actions for successfully implementing EnM and obtain enhanced EE. Regarding policy implications, the results point to the relevance of the design of environmental policy programs for effectively stimulating industrial EE.

The remainder of the paper is structured as follows. Section 2 describes the theoretical framework of the study, including a literature review of EnM practices and a presentation of the translation theory. The research methods and data collection are outlined in Section 3. Section 4 presents the empirical analysis and the results of the study. In Section 5 we offer a discussion of the research findings. Finally, in Section 6, we provide a conclusion where the limitations, implications and avenues for future research are highlighted.

2. Theoretical Framework

2.1. EnM Practices

The objective of EnM is to continuously increase manufacturing firms' EE in production processes [30]. From a wider perspective, energy management contributes to the environmental transition and sustainable development of manufacturing firms by incorporating economic and environmental factors into overall business strategies [40]. In the literature, EnM is defined as 'the strategy of meeting energy demand when and where it is needed' [48] and 'procedures for strategic work on energy [21]. Moreover, Bunse et al. [49] define 'EnM in production as including control, monitoring, and improvement activities for EE', and Ates and Durakbasa [39] emphasise that 'EnM is considered a combination of EE activities, techniques and management of related processes which result in lower energy cost and CO2 emissions'. The comprehensive definition by Schulze et al. [30] (see intro) endorses that EnM needs to be operationalised as EnM practices.

EnM practices include a broad range both technical and managerial elements [38,50], by which the spillover effect from general management practices to EE has been investigated [22]. Scholars in the field also have put forward several frameworks for characterising and assessing EnM practices in manufacturing firms, such as: Minimum requirements, Success factors for in-house EnM, EnM Matrix and Maturity models. The frameworks are described in the following text, while the EnM practices considered in these frameworks and in other relevant studies are depicted in Table 1. We identified the reviewed articles by searching for 'energy management practices' in Google Scholar and through manual screening of cross-references. The list of articles is not exhaustive but provides an overview of the EnM practices considered relevant for EE improvements in recent research.

The Minimum requirements model is a basic framework for assessing EnM in manufacturing firms [32,39], by limiting the analysis to evaluate to what extent EnM is put into practice. Based on this model, a survey study covering 304 Danish industrial firms, concluded that only between 3% and 14% of the firms practiced EnM [32]. A multiple case study in Turkey found that only 22% of the surveyed companies practiced EnM [39]. Furthermore, based on a review of empirical research on industrial EnM in Sweden, Johansson and Thollander [33] propose a framework of ten success factors for efficient in-house EnM. This framework is used in empirical studies to monitor the adoption of an

EnM program in an MNC and assesses the performance of these key elements [27]. The five level EnM matrix framework is a more elaborated framework that consists of six organisational issues related to EnM [51]. Gordic et al. [50] used this matrix when analysing the procedure for developing EnM in a Serbian car producing company. The EnM matrix is often combined with the Maturity model which is a relevant framework for evaluating firms' ability to manage energy [36,52].

The Maturity model contributes to a better understanding of suitable EnM configurations and the required steps to establish EnM practices according to firms' energy strategies [36]. Empirical studies from Sweden applying the Maturity model shows that the maturity level of firms is relatively low [36,53]. The EnM maturity and matrix models have several similarities in addressing firms' sophistication level of EnM practices and take into account more detailed descriptions of activities considered as EnM practices. The most comprehensive presentation of EnM practices is proposed by Trianni et al. [38]. Based on a literature review they develop a reference list of 58 EnM practices that are specified and can be used as a baseline for benchmarking firms' implementation of EnM practices.

Thus, the literature emphasises the importance of adopting EnM practices within an organisation and proposes assessment models with definite practices that characterise effective EnM. The framework of EnM practices depicted in Table 1 builds on this literature and serves as a point of reference when assessing the EnM practices of Pharma. In the following this is referred to as 'best EnM practice'. Table 1 synthesises the findings in the literature on how to characterise successful EnM practices. These theoretical 'best EnM practices' assert the need to adopt both management routines and organisational structures, in addition to competence-enhancing activities. Considering environmental leadership practices, it is essential that top managers support and are committed to the environmental agenda and formulate long-term environmental strategies and goals. Management practices should also focus on employee involvement and motivation. Furthermore, successful EnM practices depend on dedicated personnel working on energy matters and a clear allocation of responsibility. Additionally, performance measurement systems are essential in controlling, monitoring, and planning energy consumption against strategic targets, thus allowing for effective information assimilation and reporting to management and operational personnel. Competence is positively related to environmental awareness; hence, education and training are outlined as important ways to improve internal energy performance. Studies also assert that EnM should be reflected in a firm's investment decision processes and plans by prioritising environmental business objectives and allocating resources to EE projects. Firm characteristics and operations, such as production processes, innovation, and R&D focus are also found to influence EnM. It is worth mentioning, however, is that although some studies include energy costs and external factors as EnM practices, they are not included as such in this study since they are not considered to be part of organisational structures or routines.

Table 1. Theoretical 'best energy management (EnM) practices' based on a synthesis of scholarly articles.

Categories	Energy Management Practices (EnM Practices)	Thollander and Ottosson [8]	Martin et al. [11]	Brunke et al. [16]	Sannö et al. [27]	Schulze et al. [30]	Christoffersen et al. [32]	Johansson and Thollander [33]	Ates and Durakbasa [39]	Gordić et al. [50]	Jovanović and Filipović [52]	Sa et al. [53]
Management and Environmental leadership	Top management support and awareness of energy issues		X	X	X		X	X	X		X	
	Energy strategy (policy), planning, and targets	X	X	X	X	X		X	X	X	X	X
	Employee involvement, communication, motivations and incentives				X	X		X		X	X	X
Energy manager and Organisational structures	Energy manager and the strategic positioning of the energy manager in the organisation			X	X	X	X	X	X	X	X	X
Performance measurements	Information systems, energy audits, sub-metering, controlling and monitoring	X	X	X	X	X		X		X	X	X
Competence	Staff awareness, education, and training (culture)				X	X		X		X		X
Investment decision	Investment and pay-off criteria, and allocation of energy costs	X	X	X		X				X		X
Firm characteristics	Energy prices and competitiveness		X				X		X			
	Firm characteristics						X		X			
	Operations and production processes											
	Innovation and R & D focus		X									
External factors	Policies and regulations		X				X		X			
	External relations						X		X			

2.2. Translation Theory

In exploring the materialisation of EnM practices in Pharma, this study considers EcoFuture a management idea and analyses the firm's internal translation process of the idea. Sahlin-Andersson [43,44] defines management ideas as successful models that provide solutions to pressing problems in different contexts and at different points in time. They can further be described as social and legitimised norms for how an efficient organisation should appear regarding structural arrangements, procedures, and routines [54], such as codes of ethics [55], lean management [56] and reputation management [57]. When travelling between settings, ideas are conceived as immaterial accounts that are dis-embedded from their original contexts in terms of time, space, and location [43,57]. Hence, when an idea is re-embedded in a new setting, it must be translated and recontextualised [55]. As the translation process may include a broad range of translators, including government personnel, managing directors, middle managers, researchers, consultants, and operational personnel [58,59], there are numerous ways of translating the idea. However, empirical research has identified that translation takes place in accordance with translation or editing rules [55,56,60], that are applied more or less deliberately as a chosen strategy [47]. The outcome can be witnessed through changes in e.g., the mindset of individuals, formal documents, and the enactment of new practices [57].

Scholars have conceptualised the translation process through, for example, translation modes and rules [47,57], abstraction levels [61,62], and translation processes [63]. The theoretical framework in this study builds mainly on editing rules [43,44], which are apt for analysing the translation of broad ideas into local workplace practices. The first editing rule concerns the context and the process by which the idea is made appropriate for the local setting. In recontextualising the idea, organisational members add time, space, and sector-bounded features and make it relevant to the local setting. In this study, emphasis is placed on regulative and normative sector-bounded features and macroeconomic issues. The second editing rule concerns the formulation and labelling of an idea. The focus is on how the idea is formed so that it is deemed appropriate in the new context by discarding and adding elements to the idea [57]. Relabelling offers explanations for why an idea is successful and allows an idea to 'seem different but familiar' [56]. This rule is therefore relevant when analysing how EcoFuture was formulated when communicated in the organisation. The third rule relates to the plot of the story or the rules of logic. Sahlin-Andersson [43] describes this rule as the rationale behind the idea, in which 'explanations are given as to why a certain development has taken place'. Doorewaard and Van Bijsterveld [63] describe this as a power-based process in which the actors 'continuously reshape the elements of this process by confronting their own ideas with those of others and with existing organisational practices'. Prior empirical studies provide different examples of how to operationalise this editing rule [56,60]. Here, the analysis includes the translators' endeavours in fitting EcoFuture in the organisation setting through linking it to a known and acceptable internal logic, and thereby stimulating engagement among operational personnel. The theoretical framework is summarised in Figure 1.

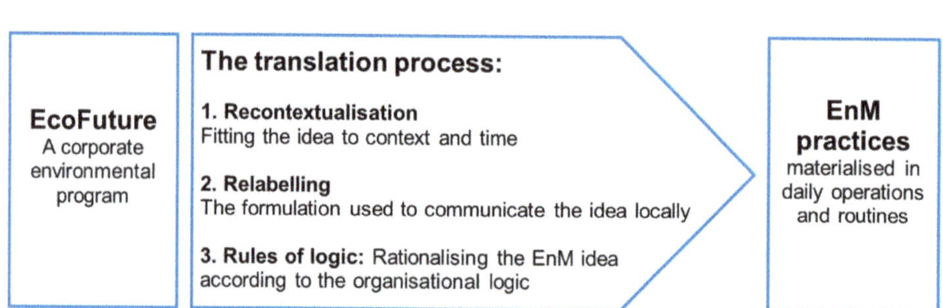

Figure 1. From program to practice: the translation of a corporate environmental program to EnM practices in a manufacturing firm.

The figure depicts how the original idea travels from the corporate to the firm level and illustrates the translation of the environmental program. It is important to point out that time is essential in the translation process, which is not captured by this model.

3. Methods

3.1. Case study Research Design

We selected a retrospective longitudinal case study as a research design when exploring the implementation of a corporate environmental program in a manufacturing firm. The case analysed here is the implementation process of a corporate environmental program in a context-bounded setting. Ragin [64] supports this understanding of a case and 'consider cases not as empirical units or theoretical categories, but as products of basic research operations'. Through the research design we aim to present a detailed understanding of the process, which can be found in qualitative data sources [65]. Furthermore, we chose the single case study approach because of its ability to provide rich and detailed data [66] on the process and the contextual setting in which the implementation occurred. Such contextual understanding is critical for the translation process and the organisational logics behind the emergence of EnM practices. Case study designs are accordingly often used in translation research [55,56,58] and in empirical studies on the implementation of environmental programs [67]. A common criticism to case study design is its inability to make statistically valid generalisations beyond the particular case. In the case of study research, however, the aim is rather to contribute to analytical generalisation [66], by analysing the contextual description of the behaviours and actions that are embedded in the empirical context through theoretical lenses [68]. Through the process of confirming, developing or extending a theory's area of use, analytical generalisation is achieved.

3.2. The Case and the Empirical Context

The case analysed in this study is the implementation process of a corporate environmental program of an MNC, here named EcoFuture. The MNC operates in a multitude of sectors, employing more than 300,000 people in over 180 countries. EcoFuture has run from 2004 and aims to integrate long-term sustainable objectives into the core of the MNC's business strategy by: (1) increasing its investment in clean R&D (cleaner technologies); (2) increasing revenue from EcoFuture products, defined as products and services that provide significant and measurable environmental performance advantages to customers; (3) reducing greenhouse gas (GHG) emissions and their intensity, along with improving EE; and (4) informing the public. At the corporate level, EcoFuture is mainly based on quantifiable environmental targets and without any detailed instructions on how it should be operationalised at the local level in the subsidiaries. The abstraction level of the program was accordingly rather high [62]. The program was globally revised in 2007, 2009, 2010, and 2014.

The empirical context selected for this study is a pharmaceutical firm in Norway with about 100 employees—Pharma. The firm specialises in producing drug substances for contrast agents used in medical imaging. Most of the firm's economic activity relates to two products, for which it holds a significant market share. The firm thus produces mainly commodity-type products of high volume for global markets, and high capital investments in facilities, which are typical in the broad-process industrial sector [40]. Pharma was purchased by the MNC in 2004 and became subject to EcoFuture. Hence, the decision to implement the program was at corporate level and not within the hands of the firm. Nonetheless, Pharma had to find means to fit the program to the local context, which is the process explored in this study.

3.3. Data Collection

The case is selected purposefully [69], due to Pharma's impressive EE improvements over time and recommendations from various sources such as environmental NGOs, energy mangers in other companies and public agencies. The case was therefore considered information-rich and adequate to answer the research question [70]. Primary data were collected through interviews, observation, firm internal manuals, and other documents during three company visits over a period of four months in 2014. Secondary data were also collected such as annual reports, conference presentations and press articles. The data collection started with a meeting with the energy manager at Pharma. At this first meeting, the overall objectives of the research project were discussed. The second company visit included a guided tour of the site and observations in some of the factories, a firm presentation, and an interview with the firm's top management. Additionally, we were given access to internal documents such as investment prospects, project descriptions, and project manuals. We also collected secondary information from press articles, marketing brochures, conference presentations, and annual corporate reports. This information was used to prepare for new interviews. During the third company visit, semi-structured interviews were carried out with key informants. The main purpose of the interviews was to understand informants' perceptions of the firm's operations, including production processes, R&D, technology implementation, and decision processes and practices for employee involvement and training. Additionally, they were asked about their understanding of EcoFuture and the integration of EnM into the firm's daily operations and activity. Following Eisenhardt and Graebner [71], data were collected from highly knowledgeable informants who viewed the focal phenomena from diverse perspectives. The energy manager helped identify key informants representing a cross-section of the organisation that were engaged in and/or affected by EcoFuture, including the director of health, safety, and environment (HSE), the energy manager, project managers, site managers, operating personnel, R&D staff, and members of the EcoFuture team. This cross-sectional sample allowed for a comprehensive understanding of the translation process and the EnM practices within Pharma. For practical reasons, some of the interviews were performed in groups of 2–3 interviewees. In total, nine key informants were interviewed. Two of the informants, considered particularly knowledgeable, were interviewed more than once. Hence, during the period, eight interviews were conducted, each lasting 1–2.5 h. All interviews were fully transcribed.

3.4. Method of Analysis

The analysis of the empirical material rests on the framework of directed content analysis [72]. The goal of a directed approach to content analysis is to validate or extend conceptually a theoretical framework or theory, and one might categorise this as a deductive approach. In our case we used translation theory, and more specifically the editing rules, as key concepts to create coding categories to guide the reconstruction of the material. As such, we also used translation theory to help focus the research question. Such theoretical conceptualisation of the empirical data is particularly suitable for case studies [70].

The analysis started with a chronological presentation of the material. This was done by triangulating data from the interviews, annual reports, conference presentations and press articles. The chronological presentation was then followed by an effort to code the material according to the editing rules described in the theoretical framework (Figure 1). Moreover, the researchers analysed the EnM practices in Pharma by using data from the interviews and observations at the site, and from internal documents and manuals describing routines and investment projects. NVivo, a qualitative analysis software product, was used to store the material and as a support for the coding. The operationalisation of the editing rules allowed the identification of shifts in intensity of the translation process that indicated patterns in the implementation of EcoFuture. Subsequently, we categorised the material into three periods: 2005–2007, 2008–2010, and 2011–2014. The first phase was mainly recognised by the contextualisation of EcoFuture to the local context of the firm. In the second phase, external economic shocks called for larger organisational changes and stimulated managerial efforts to implement EcoFuture within the organisation. The third phase was recognised by the efforts of the energy manger in rationalising EcoFuture into EnM practices. We named these phases Complacency, Urgency and Maturity, drawing the attention to some main characteristics regarding translation during these periods. As such, temporal bracketing was applied as an analytical strategy [65] by which the data are decomposed into successive periods. This strategy permitted the creation of comparative units of analysis [73] and helped us analyse how the translation of EcoFuture unfolded over time. Furthermore, this strategy allowed the inclusion of time as an additional dimension to the theoretical framework (Figure 1). Finally, the researchers assessed the outcome of the translation process, in terms of successful materialisation of EnM practices, by benchmarking Pharma's EnM practices with the theoretical best EnM practices described in Table 1.

There are some limitations related to the research design, for instance regarding the context bounded character of the data provided. This also implies challenges in making causal connections between actions and results in single case studies. Nevertheless, several strategies have been applied to ensure the rigor and reliability of the study. First, we used a research framework, derived from translation theory, to guide data collection and the analysis. Furthermore, the analysis is based on multiple data sources, and the triangulation of the material allowed us to develop an understanding of the translation process over time and reduce the intrinsic biases in empirical data.

4. Results

The presentation of results is structured according to the three chronological time periods described in Section 3.4—Complacency, Urgency and Maturity. By following the translation across these periods, the analysis illustrates how EcoFuture goes through a maturity process before unfolding as EnM practices. The results related to the translation process are summarised in Table 2 below, while the outcome of the translation process in terms of EnM practices is discussed in Section 4.4.

Table 2. Translation process of the idea according to translation rules and period.

Period	External Factors	Translation Rules			Key Translators
	Context	**Recontextualisation**	**Relabelling**	**Rules of Logic**	
	Contextual factors external to the firm affecting the translation of the idea	Fitting the idea to the context and time	Formulations used in communicating the idea locally	Rationalising the idea according to the inherent organisational logic	
Complacency	EcoFuture launched (2005) as a corporate environmental program EcoFuture targets include: Increase investment in R&D of cleaner technologies Increase revenues from products and services that provide environmental performance and advantages to customers Reduce GHG emissions and improve the EE of the firm's operations	EcoFuture was contextualised according to national environmental regulation, sector bounded technological and regulative features, in addition to limited access to capital, and translated as an EE program	-	The editors were focusing on synergies between environmental and efficiency objectives—'picking the low-hanging fruit'	Top managers
Urgency	EcoFuture is revised (2009 and 2010) with new and increased targets Other external chocks: Global financial crisis Increased global competition Establishment of an industrial network for sustainable process industry	The program was locally contextualised as an EnM and organisation development program, with focus on productivity	The program was relabelled 'Smart Growth', including emphasis on energy efficiency	Economic rationale: all EnM investments should be economically feasible	Top managers
Maturity	EcoFuture is revised (2014) with new and increased targets	-	-	Economic rationale: all energy investments should be economically feasible Rationalising the program by integrating the technological complexity of the production processes, organisational resistance to change, and existing organisational integration of energy issues	Energy manager

4.1. Period 1: Complacency

The first period is mainly recognised by the top managers' initial efforts to fit EcoFuture into the contextual setting of the firm. Normally, the translation process begins with the decision to adopt the idea [74]. In this case, however, the adoption decision was made at the corporate level, and the local managers had to find ways to fit EcoFuture to the local firm setting. Hence, the top managers contextualised EcoFuture according to regulative and technological sector-bounded features. The analysis points to the impact of pharmaceutical acts regulating lengthy procedures and product certificates before commercialising new drugs. In addition, licences and emission permits by the national environmental authorities affecting the firm's production figures also played a role in this contextualisation process. Furthermore, Pharma was characterised as having a few commodity-type products, continuous production in high volumes, extensive capital binding in existing production equipment and facilities, and limited access to investment capital. The following interviewee quote illustrates how EcoFuture imposed new environmental demands on the firm without the backing of additional resources:

> [Y]ou don't get any money for doing this. The environmental investments compete on equal terms with any other investment project. They say that we have to reduce the energy consumption, but they don't say 'Here, you have money to do it'. It doesn't work that way. The environmental projects have to enter ordinary budgets. That is tough!

These regulative and technological features prompted the managers to search for a way to implement EcoFuture without triggering significant operational changes or large investments. Subsequently, EcoFuture was translated to fit the extant portfolio of products and production processes by contextualising it as an EE program. During this first period, Pharma had significant potential for EE improvements, which were realised via minor adjustments and 'picking the low-hanging fruits'. Moreover, only the top managers were preoccupied with EcoFuture and no significant efforts were made to implement the program within the entire organisation. Such situations in which an idea resides high in the hierarchy being decoupled from organisational practice, is described as 'isolation' by Røvik [74], and casued the further travel of EcoFuture to stagnate. The analysis suggests a complacent translation of EcoFuture during this period, in the sense that translators in Pharma did not see the relevance or need to take large environmental steps and putting efforts into translating EcoFuture.

4.2. Period 2: Urgency

This second period is recognised by external economic shocks and changes in the institutional environment that boosted the relevance of EcoFuture. With this as a backdrop, the translation of the program intensified, as the top managers started to recontextualise and relabel EcoFuture with the aim to implement the program within the organisation. Indeed, the global financial crisis of 2008 led to reduced sales, price drops, and production overcapacity. Moreover, the patent protection on Pharma's products expired during this period, leading to increased competition from generic drug manufacturers. EcoFuture was also revised at the corporate level and the environmental targets amplified. As the firm had already taken advantage of the 'low-hanging fruits', greater organisational involvement was now needed, and it became necessary to rethink the strategy and organising of the program.

Economic considerations were prominent when top managers translated the program during this period. Furthermore, in translating EcoFuture into an internally understandable concept, the program was relabelled as a productivity program named 'Smart Growth', with an emphasis on EnM. The use of familiar rhetoric is an efficient means to avoid organisational resistance towards an idea [55]. Productivity was a well-established concept within the organisation. EcoFuture and EnM were, hence, fitted to the local setting by having productivity as a primary objective and taking advantage of the synergies between increased productivity and EE. This approach is illustrated by an interviewee stating that the 'environmental benefits are a spin-off of productivity'.

The intensification of the translation process also involved the restructuring of the organisation and technical improvements. Furthermore, new EnM practices started to materialise such as energy audits, the systematic monitoring of main energy streams, and the measurement of total energy consumption, all of which are essential for attaining correct information, reporting, transparency, and constructing a shared course of action in strategic energy planning [75]. In addition, the operational personnel started to engage in EE. As illustrated by the following interviewee quote, this upscaling of the program can be considered an important premise for succeeding with energy improvements:

> *[F]rom then, the program changed from being an energy-saving program that only some were engaged in to become a factory productivity program. This is one of the success criteria.*

Contextual changes characterised this period. The analysis indicates that these changes created an urge to act in the organisation [74] that reactivated the relevance of EcoFuture. Hence, in contrast to the complacency described in the first period, the idea was now translated as an EnM and productivity program. As a result, the initially abstract idea now started to materialise further in the organisation. This could be observed in both the relabelling of EcoFuture and the emergence of EnM practices.

4.3. Period 3: Maturity

During the first two periods, EcoFuture was translated into an EnM program after being recontextualised and relabelled primarily by the top managers. In contrast, this third period is recognised by the further materialisation of new EnM practices. The analysis now reveals the energy manager as an important translator. This includes his efforts in aligning EnM with internal organisational agendas and structures, such as the technological complexity of the production processes, organisational resistance to change, organisational integration of energy issues, and the rationale of economic feasibility. By applying 'rules of logic' as a translation tool, the energy manager stimulated the development of accepted and legitimate EnM practices.

First, we find that the rationale of economic feasibility was used as a way of framing the implementation of the EnM program. In general, investments arise due to new technology, machinery wear-out, and changes in market demand, prices, and legal requirements. Investments are, thus, necessary to remain competitive in changing environments. The analysis indicates that the principle of 'good business' was an overarching logic in Pharma and thereby strongly affecting the internal investment decision processes and reporting schemes. Good business is understood as investments that are economically feasible in the short term and ideally lead to long-term sustainability. Indeed, all investments were based on economic considerations within which environmental improvements were deemed positive spin-offs. This approach is exemplified in the following interviewee quote:

> *We don't really think that our first priority is to save the environment. However, we include it in productivity. We will show you how we plan to become CO2 neutral ... It is, however, not a target in itself for this factory, even though there are some demands for us to become CO2 neutral.*

The significance of the economic logic is also apparent in the way investment projects were evaluated and prioritised. The typical required payback time for investments was 2–3 years. Prior studies have noted that it is a challenge for energy projects to comply with such short payback demands, thus causing a significant barrier for industrial EE improvements [76,77]. Furthermore, all investment projects were categorised and rated according to compliance, health, safety and environment (HSE), maintenance, and productivity. The following interviewee quote illustrates the limited priority given to environmental objectives:

> *Legislative compliance, HSE, and maintenance have priority before the green projects (because the green projects are sometimes productivity-related), and we don't get green projects through just because they are green.*

The firm had organisational structures and routines for assimilating information and reporting to management and the organisation about energy consumption. Nonetheless, to gain support for energy-related projects during investment decision processes, energy improvements had to be argued and rationalised according to the extant economic logic and culture of the firm. Hence, arguments based on compliance, HSE, maintenance, and productivity had to be integrated and highlighted to win through with environmental projects. These strategies of searching for and aligning economic and environmental objectives are important for rationalising the EnM program according to the organisational logic. Nonetheless, these strategies contradict with recommendations from prior studies asserting that earmarked funding for environmental investments is a significant driver for EE [76]. Moreover, Sandberg and Söderström [78] state that the priorities made during investment decision processes depend on the culture of the firm. In other words, the strong economic rationale can be considered a cultural factor affecting the rationalisation of the EnM program in Pharma.

In this situation, the energy manager had seniority and operational experience, which provided him with in-depth knowledge about the factory and production processes. The analysis suggests that he took advantage of this experience and previously established energy monitoring practices -which provided him with accurate information on energy consumption, production bottlenecks, and investment opportunities—when rationalising the EnM program according to the logic of 'economic feasibility'. Radaelli and Sitton-Kent [59] describe several channels that middle managers can use in this translation process. Here, the energy manager worked proactively and used formal and informal arenas to continuously sell the EnM idea, manage conversations, and align diverging interests among organisational units and members.

Pharma had developed a complex production infrastructure over decades. Any changes in the production processes, including energy improvements, required not only excellent engineering skills, but also comprehensive and in-depth knowledge about the factory. External stakeholders were considered by Pharma to lack such detailed knowledge; thus, the innovation processes in the firm depended on the skills and creativity of the employees. Consequently, Pharma's EnM practices were largely determined by the employees' motivation and competencies in EnM. This finding coincides with prior studies underlining the role of operational personnel in process innovation and EnM [25] and the significance of employee motivation regarding EnM [31]. Hence, in addition to technological complexity, the translation of a management ideas depends on translators' ability to mitigate organisational resistance to change [60]. Such resistance was also experienced in Pharma, with the following interviewee quote illustrating the challenge of motivating employees to work collectively with EnM:

> [W]e have some examples [in which] we have completed projects in one area where the savings have been in another area, and then we have met a lot of resistance in the area in which the change has taken place.

The integration of energy issues in organisational structures, or lack of such integration, was also part of the translation process. Pharma had systems and routines for controlling and monitoring the largest energy flows, however the analysis indicates that energy issues were only formally integrated in selected areas of the organisation. This restricted integration can be exemplified by the firm's use of key performance indicators (KPIs). KPI is a management tool developed to motivate employees in a preferred direction and is a recommended EnM practice [33]. Nonetheless, the use of KPI can also hamper change processes and compromise EnM practices [24], by stimulating other operational activities and investments. In Pharma, KPI was customised to the main activity of each unit. Hence, only the units with extensive energy consumption had energy use integrated into their KPI. Consequently, this use of KPI provided most employees with limited incentives for engaging with the firm's total energy consumption.

However, to motivate operational personnel to work collectively with EnM, Pharma provided several competence-enhancing schemes and activities such as internal training programs and education support, which are found to be significant drivers for industrial EE [28,79]. The following interviewee quote illustrates how the benefits of energy improvements at the individual level were identified and communicated to provide motivation to make energy improvements:

> *The best projects are carried out when the area in which the change takes place [obtains] good benefits from the change. So, being very strategically smart, when you create projects, you have to find something that the department [in which] the change [takes place] benefits from.*

The energy manger also formed alliances with external stakeholders. The use of an external network is a well-known strategy for enhancing EE by exchanging knowledge and ideas [80]. Pharma was a partner in an industrial network for sustainable process industry firms. This network in addition to other external stakeholders were used strategically to gain legitimacy for the program and convey the relevance of EnM to the organisation by promoting Pharma's EnM at conferences, generating editorial publicity, and inviting firms, experts, academics, non-governmental organisations, and politicians to the site.

Furthermore, the energy manager's central position in the organisation allowed him to communicate directly to top management and have close relationships with operational personnel. The energy manager's position in the organisation is known to affect EnM implementation [11,40]. Personal relationships are also known to be valuable assets in innovation processes [81]. Here, the energy manager took advantage of his position in the organisation to get involved in the conceptual design of projects early on. The following quote by the energy manager illustrates how such early involvement is essential for aligning the demands, interests, and rationales of other organisational members:

> *Here, we have a project that is very good. However, it does not include that much energy saving. It is a project that is about reducing solvent in one area. It provides a yield increase and some energy savings … I think there is more energy to be saved! Since it includes energy saving, I have worked with the concept.*

By getting involved early in the conceptual design of new projects, the energy manager uses his expertise when strategically aligning the EnM program with the organisational rationale, engaging allies, and inviting opponents to identify common goals and values. Accordingly, the analysis indicates the relevance of timing and early involvement in the translation process.

The analysis suggests that the process of translating EcoFuture into EnM practices involves selling a version of the idea, mediating others' versions, and aligning goals and agendas, thus supporting Doorewaard and Van Bijsterveld [63] who describe translation as a power-based process in which the involved actors 'continuously reshape the element of this process by confronting their own ideas with those of others and with existing organisational practices'. Furthermore, this period is characterised by the energy manager's efforts to rationalise the EnM program. Although top managers' involvement as translators was significant in the first two periods, the energy manager appeared as a key translator during this third period. The results described in this section reflects that Pharma has now reached a higher level of maturity [27,51] regarding the implementation of the EnM program and the EnM practices. Table 2 depicts the implementation process of EcoFuture in terms of translation rules and periods. The results suggest that the expected chronology in the use of the translation rules, as depicted in the conceptual framework, was not met. Instead the use of the rules had a more dynamic character. Rather than applying one translation rule in each period, the process is characterised by successive phases of translation. Within each phase, various translation rules with different intensities were applied, and different translators at various levels in the organisation were involved. We label this pattern as translation dynamics.

4.4. EnM Practices in Pharma

The results of the translation process are evaluated based on the premise of the materialisation of EnM practices and with reference to the theoretical 'best EnM practices' (Table 1). The identified EnM practices in Pharma are listed in Table 3. Inspired by the EnM maturity models [27,51], we have separated into two columns the EnM practices in Pharma that are considered to comply with the best EnM practices, from those considered not-complying.

The results suggest that the EnM practices in Pharma to a large extent comply with the theoretical best EnM practice. Indeed, several EnM practices related to management, organisational routines and structures, and competences are identified in the firm. However, the analysis also reveals shortcomings in some of Pharma's EnM practices that the literature emphasises as essential. The first shortcoming concerns Pharma's use of KPIs, which for most units in Pharma do not include energy consumption or other energy related measures. Indeed, while most good management practices have beneficial spillovers on EE, an emphasis on non-energy targets is found to be correlated with energy inefficiency [24]. Hence, Pharma's design and use of KPIs might undermine the employees' motivation to improve the firm's overall EE. The second shortcoming concerns Pharma's limited integration of environmental strategies in the internal investment decision processes. Such decoupling between energy objectives and firms' investment decision processes is known as a considerable barrier for EE improvement [76,77]. Nevertheless, Pharma improved its EE considerably over the analysed period, which might seem to be a paradox. Some possible explanations will be discussed in the following section.

Table 3. Assessing the EnM practices in Pharma with reference to the theoretical 'best EnM practice'.

		EnM Practices in Pharma	
Cat.	Best EnM Practises	Complying Practices	Not-Complying Practices
Management and Environmental leadership	Top management support and awareness of energy issues	Top management awareness of and support for EnM program Top managers are aware of and support environmental issues and the value of energy savings	
	Energy strategy (policy), planning, and targets	Long- and short-term energy reduction targets	
	Employee involvement, communication, motivations and incentives	Employee involvement Routines to provide information about energy consumption to management and the organisation	Only production areas with a high consumption of energy have energy-related KPIs and are, hence, motivated to engage in energy-saving projects; other areas lack the same incentives
Energy manager and Organisational structures	Energy manager and the strategic positioning of the energy manager in the organisation	Allocation of resources to energy issues; two full-time employees (including the energy manager) are working with environmental reporting and energy-saving projects, with the ability to involve others when needed Energy manager is positioned strategically in the organisation and reports directly to the top management	
Performance measurements	Information systems, energy audits, sub-metering, controlling and monitoring	Energy audits Systematic monitoring and measuring of energy consumption Systems and routines for controlling and monitoring the largest energy flows	
Competence	Staff awareness, education, and training (culture)	Internal and external EnM-related training and education programs Culture of employee involvement, communication, and cooperation between departments to identify good technological solutions for reducing energy consumption	
Investment decision	Investment and pay-off criteria, and allocation of energy costs		No earmarked investment capital for EnM; EnM projects are evaluated similar to all other projects according to compliance, HSE, maintenance, and productivity. Investments in energy projects are assessed according to a short payback time of 2–3 years

5. Discussion

Through the lens of the editing rules, as part of the translation theory, a picture of the implementation process of EcoFuture—from program to practice— emerges from the results (Section 4), illustrated in Figure 2 below:

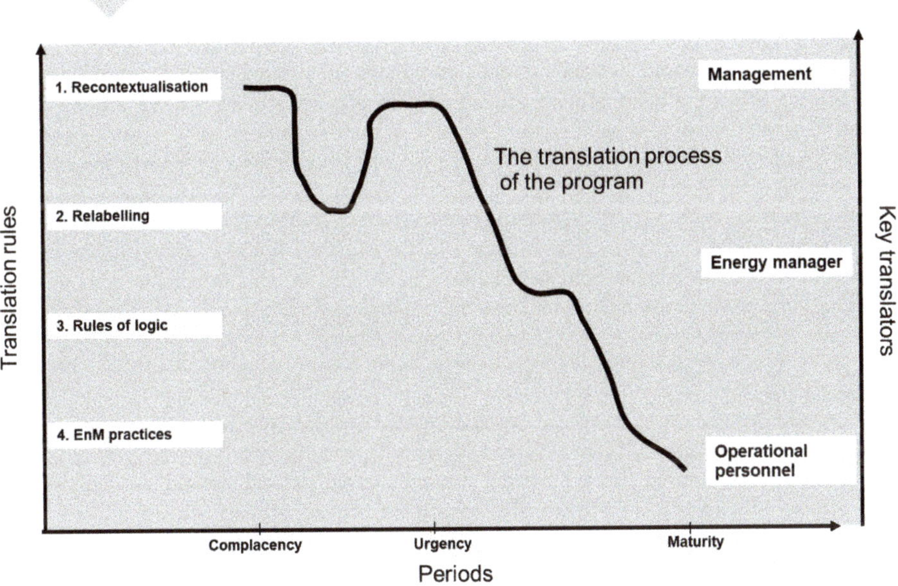

Figure 2. Conceptual model of the EnM implementation process from program to practise.

Figure 2 conceptualises the translation process and the resultant EnM practices in the studied firm. The figure incorporates three essential dimensions of the implementation process, namely: time, translation rules and key translators. The translation of the corporate environmental program passed through three time periods, which were recognised by changes in translation intensity, and the activation of different translators. The figure also illustrates how the translation moved back and forth over a decade, driven by different translation rules and key translators, before resulting in new EnM practices.

The results suggest that the use of the translation rules has a dynamic character. Rather than applying the translation rules successively in each period in a linear manner, the periods are recognised by different dynamics of translation. Within each period, various translation rules were applied with different intensity, and with the involvement of different translators. This finding supports prior studies stating that ideas need to undergo several cycles of translation before being applied to a new setting [56]. It also coincides with Røvik's [74] remark that ideas might alternate between passive and active phases and may linger in an organisation for a long time before materialising, leading to a gradual, slow-phased transformation of the idea to practice. Interestingly, this pattern of translation dynamics that stretches over multiple years, is an aspect that is not pinpointed in the EnM literature. Within this field, EnM implementation is usually considered from a static perspective, focusing on models and tools for assessing the end results in terms of EnM practices [38]. Moreover, the efforts and competences required to obtain to these results have received little attention [25,30]. Hence, this study complements extant research by illustrating the complexity of implementation, and emphasising the relevance of long-time perspectives and the endurance of managers and other key translators from different levels in the organisation.

Following the analysis, we see how the top managers played an important role during the first two periods of the process. However, the energy manager emerged as a key translator when rationalising the program into EnM practices that were accepted and adopted by the operational personnel. Energy managers are often middle managers that both lack the hierarchical authority of top managers, and the immediate operational knowledge of operational personnel [59]. It is however confirmed that energy managers' operational experience [77] and position in the organisational hierarchy [11] have a positive effect on firms' EE. In line with this research, our study addresses how the energy manager used his operational experience, technical competence, social network and position in the organisation to champion the environmental program. Indeed, the results show how the energy manger worked actively, using formal and informal arenas, to rationalise energy improvements using a logic that was deemed legitimate and agreed upon in the organisation. In this manner the energy manger took the role as a change agent that 'made the change happen' [82], being translators both of the idea to be implemented and in terms of 'translating' between different levels in the organization due to their position as two-way-windows [83]. Coinciding with the increasing recognition of middle managers' role in change processes [84], our finding supports the relevance of energy managers to drive the environmental transition in manufacturing firms.

In addition to the findings that we have highlighted in Figure 2, there are two other points to be made as part of the discussion. One concerns the abstraction level of the idea, and another is the identification of additional EnM practices to those being part of the theoretical 'best EnM practices' (Table 1).

Translation scholars hold that there are variations in the abstraction level of management ideas. The abstraction level reflects whether the idea contains explicit and detailed descriptions of how it should be materialised into practices, or whether it gives room for the local organisation to make its own interpretation of the contents [62]. It is further argued that abstract ideas are more complex and thus harder to implement in a new context than those that have a more explicit and concrete character [47]. In this case, EcoFuture was oriented towards quantifiable targets without providing details on how to operationalise the program locally and is thus considered to hold a high abstraction level. The results show that the abstraction level of the environmental program allowed the translators large flexibility and freedom in the translation process, and that EcoFuture changed quite extensively when fitted to the local firm setting. In contrast to Røvik' suggestion [47], our findings suggest that the high abstraction level of the program was a success criterion rather than a barrier for the implementation of the program. As such, this finding contributes to shed some light on the challenges that firms face when they are going to implement EnM standards. A key argument in this study is that the abstract character of the environmental program implies that its contents are negotiable in each new setting. Furthermore, we argue that each local translation of the program can give rise to new versions and result in differing practices that is customised to the contextual setting of the individual firm.

The results show that most of the EnM practices recommended in the literature were put into action in Pharma. As such, the study complements prior research in emphasising the relevance of EnM practices related to environmental leadership, organisational structures and routines, and competence-enhancing activities. However, compared to the theoretical best EnM practices (Table 1), there are some vital shortages in the Pharma's EnM practices. In particular, this relates to the investment decision processes and the use of KPIs, under which environmental issues are granted limited priority. Financial limitations and firms' reluctance to prioritise EnM at a strategic level are considered substantial barriers to EE [8,40]. Despite this, Pharma reports exemplary records of EE improvements during the analysed period. Although the context bounded character of the data provided have limitations and must be closely considered, the results point to an interesting finding which we suggest is not properly captured in previous EnM studies, and which might provide an explanation to this apparent paradox. By translating EnM according to the extant economic logic embedded in the organisation, EnM gained legitimacy and strategic relevance. Investment projects were accordingly conceptualised so that both economic and energy objectives were attained. In this way, the firm surmounted financial and organisational barriers and attained continuous EE improvements. This finding underlines the importance of translators of the environmental program working actively to align environmental and economic objectives in projects, and thereby gaining support in investment decision processes, even without earmarking funding for environmental projects. Moreover, this finding supports the claim that management ideas' capacity to travel depend on the extent to which they are associated with rational values such as renewal, efficiency, and effectiveness [85]. This way of using editing rules in translation processes invites the proposal of an additional EnM practice to the list of 'best practice EnM', one called 'translation competence' [47]. Translation competence points to the ability and knowledge of key actors and change agents to translate ideas and programs between organisational contexts. The production and diffusion of organisational ideas are important tools to transfer best practices between organizations [55,58]. Since the use of translation competence is assumed to increase the probability of achieving the desired organisational ends [43,46,47], this competence becomes increasingly important as a strategic organizational resource [47].

The current discourse in the field emphasises EnM models, and in particular maturity models [51], as efficient tools for reaching enhanced industrial EE [25,38,52,53]. These models recommend explicit EnM practices and consider them as the basic element of the analysis of EnM activity in firms [38], commonly rated as maturity levels [51,52]. Hence, the scope of these models is on the preferred end result. What is missing though in these studies, and in the broader literature, is a conceptual understanding and awareness of firms' internal processes that lead to this organisational end. In this study we demonstrate that successful implementation of EnM, in terms of the materialisation of EnM practices, is intimately connected to the organisational acceptance and preparation of the program. Furthermore, we establish the potential of the translation framework in research on EnM, by illustrating the relevance of translation rules [43,44] as a tool to guide successful implementation of EnM. Moreover, the study supports an orientation towards 'good translations' as part of these processes [44,45,59], to increase the probability of achieving more mature levels of EnM practices in the firm. Through the concept of good translation, attention is directed towards 'translation competence' in terms of managers' awareness of translation rules, abstraction level of ideas and the role of key translators. Consequently, while scholars have addressed the need for a best EnM practice [30] and EnM practice-based assessment models [38], we here propose the relevance of a 'translation framework' that complement these assessment models by supporting managers during the acutal implementaion process of environmental programs. As such, we propose a way of complementing 'result' with 'process', which can serve as reference point for both management and policy implications.

Regarding management implications, the study emphasises that organisations play an active role in translating how environmental programs are operationalised and that these processes are lengthy and often last for years. Thus, successful implementation requires managerial endurance, support, and dedication. Moreover, we suggest translation competence as part of these processes [45,46], as it may increase the probability of achieving organisational ends [47]. Managers should accordingly direct attention to good translations in terms of being aware of translation rules, the abstraction level of the idea and key translators. Furthermore, as middle managers and other employees play a prominent role during this process, managers need to encourage and educate individuals and set up organisational structures supporting the environmental change process.

Considering policy recommendation, the study suggests a positive relationship between EnM and EE in manufacturing firms and indicates that EnM is an important means to attain environmental targets. To enhance EnM practices in manufacturing firms, the results suggest that policymakers consider three mechanisms: regulation, idea complexity, and education. Policies and regulations on energy consumption and GHG emissions require organisations to adopt a concept or maintain a certain practice, as legislative compliance is a precondition for business operation [86]. Voluntary agreements at local levels of governments are also effective in overcoming the traditional constraints of implementing top-down policies at the local level [37], and allow each firm to identify the solutions that are deemed most fitted for the local setting. The study also illustrated how the abstraction level of an idea affects the flexibility in how the environmental idea is implemented at firm level. An abstraction level that is either too high or too low might pose challenges for implementation [62]. ISO-standards [31] can for example be characterised as explicit and detailed content-wise and therefore with a low abstraction level. Hence, policymakers must be observant and ensure that the policy framework contains all relevant information required to explain and understand the EnM practices and be flexible enough to fit them into the local setting. Furthermore, this study shows that the emergence of EnM practices depends on the competence of the translators and the amount of resources devoted to educating and training organisational members. This implies that energy policies should support EnM-related education and on-the-job training.

6. Conclusions

This study explores the implementation process of a corporate environmental program in a manufacturing firm, a perspective that has received limited attention in the EnM literature. The conceptual framework is built on translation theory and the EnM literature, and the qualitative analysis is based on data from a pharmaceutical firm over the period 2004–2014. The analysis suggests that a wide spectrum of EnM practices have materialised during the studied period, and complements prior research suggesting a positive link between EnM practices and EE in manufacturing firms. Furthermore, this study addressed the relationship between the translation process of a corporate environmental program and the successful materialisation of EnM practices.

The results point to four main findings with theoretical relevance to the field. First, the pattern of translation dynamics illustrates how the implementation process consists of various periods of translation that evolve over time. Second, the results point to the role of the energy manager as a key translator when rationalising the idea into EnM practices. Third, the relevance of the abstraction level of management ideas is addressed. Fourth, we propose a new EnM practice to the list of 'best practice EnM', namely 'translation competence'.

Based on these findings, the study proposes avenues for future research. The translation occurs in a dynamic environment in which both the idea and context change over time [58]. Hence, it is difficult to determine whether the EnM practices emerged as a direct result of the idea or whether they would have emerged regardless of the adaptation. More research is therefore needed to obtain more knowledge on this relationship. Moreover, although this study focused on the translation process in a recipient organisation, little is known about how to effectively prepare an idea for new settings. Hence, there is need for more research about the decontextualisation phase of environmental programs—that is, translating the desired practices into an abstract representation (e.g., images, words, and texts) that is easy to recontextualise at the firm level. Such knowledge can give rise to valuable recommendations to policymakers on how to design environmental policy frameworks that can easily travel across contexts and organisations.

Furthermore, even though this study demonstrates the potential of translation theory in research on EnM, we have only applied some selected elements from this rather mature framework which has gained attention and attracted different fields [41]. Due to the pressing relevance of sustainable management programs and sustainable business models, the development and diffusion of organizational ideas has become increasingly important for businesses as a strategic organizational resource, for research and for the larger society. Hence, we suggest that EnM scholars should explore further the potential within this theoretical framework to further conceptualise translation competence as an EnM practice and develop best translation models of EnM.

Author Contributions: Writing—original draft: M.T.S.; writing: review & editing: M.T.S. and E.A.N. Both authors have read and agreed to the published version of the manuscript.

Funding: This research received no external funding.

Acknowledgments: The publication costs for the article are funded by the publication fund of UiT The Arctic University of Norway.

Conflicts of Interest: The authors declare no conflict of interest.

References

1. IEA. World Energy Balances: Overview. 2018. Available online: https://webstore.iea.org/download/direct/2263?fileName=World_Energy_Balances_2018_Overview.pdf (accessed on 6 August 2019).
2. Brundtland, G.H. *Our Common Future*; UN World Commission on Environment and Development, Oxford University Press: Oxford, UK, 1987.
3. UNFCCC. Adoption of the Paris Agreement. 2015. Available online: https://unfccc.int/resource/docs/2015/cop21/eng/l09r01.pdf (accessed on 18 February 2018).
4. EU. *2030 Climate & Energy Framework*; Com (2015) 15 final; European Union: Brussels, Belgium, 2014; Available online: http://eur-lex.europa.eu/legal-content/EN/TXT/PDF/?uri=CELEX:52014DC0015&from=EN (accessed on 18 February 2018).
5. Trianni, A.; Cagno, E.; De Donatis, A. A framework to characterize energy efficiency measures. *Appl. Energy* **2014**, *118*, 207–220. [CrossRef]
6. EIA. Energy Efficiency and Conservation. 2017. Available online: http://www.eia.gov/energyexplained/index.cfm?page=about_energy_efficiency (accessed on 18 February 2018).
7. Backlund, S.; Thollander, P.; Palm, J.; Ottosson, M. Extending the Energy Efficiency Gap. *Energy Policy* **2012**, *51*, 392–396. [CrossRef]
8. Thollander, P.; Ottosson, M. Energy Management Practices in Swedish Energy-Intensive Industries. *J. Clean. Prod.* **2010**, *18*, 1125–1133. [CrossRef]
9. Costa-Campi, M.T.; García-Quevedo, J.; Segarra, A. Energy efficiency determinants: An empirical analysis of spanish innovative firms. *Energy Policy* **2015**, *83*, 229–239. [CrossRef]
10. Martínez, C.I.P. Investments and energy efficiency in colombian manufacturing industries. *Energy Environ.* **2010**, *21*, 545–562. [CrossRef]
11. Martin, R.; Muûls, M.; De Preux, L.B.; Wagner, U.J. Anatomy of a Paradox: Management Practices, Organizational Structure and Energy Efficiency. *J. Environ. Econ. Manag.* **2012**, *63*, 208–223. [CrossRef]
12. Fan, L.W.; Pan, S.J.; Liu, G.Q.; Zhou, P. Does energy efficiency affect financial performance? Evidence from chinese energy-intensive firms. *J.Clean. Prod.* **2017**, *151*, 53–59. [CrossRef]
13. Worrell, E.; Laitner, J.A.; Ruth, M.; Finman, H. Productivity benefits of industrial energy efficiency measures. *Energy* **2003**, *28*, 1081–1098. [CrossRef]
14. Cagno, E.; Trianni, A. Exploring Drivers for Energy Efficiency within Small- and Medium-Sized Enterprises: First Evidences from Italian Manufacturing Enterprises. *Appl. Energy* **2013**, *104*, 276–285. [CrossRef]
15. Sorrell, S.; Mallett, A.; Nye, S. *Barriers to Industrial Energy Efficiency: A Literature Review*; Working Paper 10/2011 UNIDO; United Nations Industrial Development Organization: Vienna, Austria, 2011.
16. Brunke, J.-C.; Johansson, M.; Thollander, P. Empirical Investigation of Barriers and Drivers to the Adoption of Energy Conservation Measures, Energy Management Practices and Energy Services in the Swedish Iron and Steel Industry. *J. Clean. Prod.* **2014**, *84*, 509–525. [CrossRef]
17. Cui, Q.; Li, Y. An empirical study on energy efficiency improving capacity: The case of fifteen countries. *Energy Effic.* **2015**, *8*, 1049–1062. [CrossRef]
18. IEA. *World Energy Outlook*; Internatinal Energy Agency: Paris, France, 2012; Available online: http://www.iea.org/publications/freepublications/publication/English.pdf (accessed on 15 July 2017).
19. Lin, B.; Tan, R. Ecological total-factor energy efficiency of china's energy intensive industries. *Ecol. Indic.* **2016**, *70*, 480–497. [CrossRef]
20. Jaffe, A.B.; Stavins, R.N. The energy-efficiency gap What Does It Mean? *Energy Policy* **1994**, *22*, 804–810. [CrossRef]
21. Lawrence, A.; Thollander, P.; Karlsson, M. Drivers, Barriers, and Success Factors for Improving Energy Management in the Pulp and Paper Industry. *Sustainability* **2018**, *10*, 1851. [CrossRef]

22. Cheng, Y.; Awan, U.; Ahmad, S.; Tan, Z. How do technological innovation and fiscal decentralization affect the environment? A story of the fourth industrial revolution and sustainable growth. *Technol. Forecast. Soc. Chang.* **2021**, *162*, 120398. [CrossRef]

23. Awan, U.; Imran, N.; Munir, G. Sustainable development through energy management: Issues and priorities in energy savings. *Res. J. Appl. Sci. Eng. Technol.* **2014**, *7*, 424–429. [CrossRef]

24. Boyd, G.A.; Curtis, E.M. Evidence of an 'Energy-Management Gap' in U.S. Manufacturing: Spillovers from Firm Management Practices to Energy Efficiency. *J. Environ. Econ. Manag.* **2014**, *68*, 463–479. [CrossRef]

25. Lawrence, A.; Nehler, T.; Andersson, E.; Karlsson, M.; Thollander, P. Drivers, Barriers and Success Factors for Energy Management in the Swedish Pulp and Paper Industry. *J. Clean. Prod.* **2019**, *223*, 67–82. [CrossRef]

26. May, G.; Stahl, B.; Taisch, M.; Kiritsis, D. Energy Management in Manufacturing: From Literature Review to a Conceptual Framework. *J. Clean. Prod.* **2017**, *167*, 1464–1489. [CrossRef]

27. Sannö, A.; Johansson, M.T.; Thollander, P.; Wollin, J.; Sjögren, B. Approaching Sustainable Energy Management Operations in a Multinational Industrial Corporation. *Sustainability* **2019**, *11*, 754. [CrossRef]

28. Solnørdal, M.; Foss, L. Closing the Energy Efficiency Gap—A Systematic Review of Empirical Articles on Drivers to Energy Efficiency in Manufacturing Firms. *Energies* **2018**, *11*, 518. [CrossRef]

29. Cagno, E.; Trianni, A.; Spallina, G.; Marchesani, F. Drivers for energy efficiency and their effect on barriers: Empirical evidence from Italian manufacturing enterprises. *Energy Effic.* **2017**, *10*, 855–869. [CrossRef]

30. Schulze, M.; Nehler, H.; Ottosson, M.; Thollander, P. Energy Management in Industry—A Systematic Review of Previous Findings and an Integrative Conceptual Framework. *J. Clean. Prod.* **2016**, *112*, 3692–3708. [CrossRef]

31. Virtanen, T.; Tuomaala, M.; Pentti, E. Energy Efficiency Complexities: A Technical and Managerial Investigation. *Manag. Account. Res.* **2013**, *24*, 401–416. [CrossRef]

32. Christoffersen, L.B.; Larsen, A.; Togeby, M. Empirical Analysis of Energy Management in Danish Industry. *J. Clean. Prod.* **2006**, *14*, 516–526. [CrossRef]

33. Johansson, M.T.; Thollander, P. A Review of Barriers to and Driving Forces for Improved Energy Efficiency in Swedish Industry—Recommendations for Successful In-House Energy Management. *Renew. Sustain. Energy Rev.* **2018**, *82*, 618–628. [CrossRef]

34. Thollander, P.; Palm, J. Industrial Energy Management Decision Making for Improved Energy Efficiency—Strategic System Perspectives and Situated Action in Combination. *Energies* **2015**, *8*, 5694–5703. [CrossRef]

35. ISO. The ISO Survey of Management System Standard Certifications 2016. Available online: https://isotc.iso.org/livelink/livelink?func=ll&objId=18808772&objAction=browse&viewType=1 (accessed on 18 February 2018).

36. Sa, A.; Thollander, P.; Cagno, E. Assessing the driving factors for energy management program adoption. *Renew. Sustain. Energy Rev.* **2017**, *74*, 538–547. [CrossRef]

37. Eichhorst, U.; Bongardt, D. Towards Cooperative Policy Approaches in China–Drivers for Voluntary Agreements on Industrial Energy Efficiency in Nanjing. *Energy Policy* **2009**, *37*, 1855–1865. [CrossRef]

38. Trianni, A.; Cagno, E.; Bertolotti, M.; Thollander, P.; Andersson, E. Energy management: A practice-based assessment model. *Appl. Energy* **2019**, *235*, 1614–1636. [CrossRef]

39. Ates, S.A.; Durakbasa, N.M. Evaluation of Corporate Energy Management Practices of Energy Intensive Industries in Turkey. *Energy* **2012**, *45*, 81–91. [CrossRef]

40. Rudberg, M.; Waldemarsson, M.; Lidestam, H. Strategic Perspectives on Energy Management: A Case Study in the Process Industry. *Appl. Energy* **2013**, *104*, 487–496. [CrossRef]

41. Wæraas, A.; Nielsen, J.A. Translation theory 'translated': Three perspectives on translation in organizational research. *Int. J. Manag. Rev.* **2016**, *18*, 236–270. [CrossRef]

42. Czarniawska, B.; Joerges, B. Travels of ideas. In *Translating Organizational Change*; Czarniawska, B., Sevón, G., Eds.; Walter de Gruyter: Berlin, Germany, 1996.

43. Sahlin-Andersson, K. Imitating by editing success: The construction of organizational fields. In *Translating Organizational Change*; Czarniawska-Joerges, B., Sevón, G., Eds.; Walter de Gruyter: Berlin, Germany, 1996; pp. 69–92.

44. Sahlin, K.; Wedlin, L. Circulating ideas: Imitation, translation and editing. In *Handbook of Organisational Institutionalism*; Greenwood, R., Oliver, C., Sahlin, K., Suddaby, R., Eds.; Sage: London, UK, 2008; pp. 218–242.

45. Øygarden, O.; Mikkelsen, A. Readiness for Change and Good Translations. *J. Chang. Manag.* **2020**, 1–27. [CrossRef]

46. Nilsen, E.A.; Sandaunet, A.G. Implementing new practice: The roles of translation, progression and reflection. *J. Chang. Manag.* **2020**. [CrossRef]

47. Røvik, K.A. Knowledge Transfer as Translation: Review and Elements of an Instrumental Theory. *Int. J. Manag. Rev.* **2016**, *18*, 290–310. [CrossRef]

48. Abdelaziz, E.A.; Saidur, R.; Mekhilef, S. A review on energy saving strategies in industrial sector. *Renew. Sustain. Energy Rev.* **2011**, *15*, 150–168. [CrossRef]

49. Bunse, K.; Vodicka, M.; Schönsleben, P.; Brülhart, M.; Ernst, F.O. Integrating energy efficiency performance in production management—Gap analysis between industrial needs and scientific literature. *J. Clean. Prod.* **2011**, *19*, 667–679. [CrossRef]

50. Gordić, D.; Babić, M.; Jovičić, N.; Šušteršič, V.; Končalović, D.; Jelić, D. Development of Energy Management System—Case Study of Serbian Car Manufacturer. *Energy Convers. Manag.* **2010**, *51*, 2783–2790. [CrossRef]

51. Carbon Trust. The Energy Management Guide CTG054. Available online: https://www.carbontrust.com/energymanagement (accessed on 15 October 2020).

52. Jovanović, B.; Filipović, J. ISO 50001 standard-based energy management maturity model—Proposal and validation in industry. *J. Clean. Prod.* **2016**, *112*, 2744–2755. [CrossRef]

53. Sa, A.; Thollander, P.; Cagno, E.; Rafiee, M. Assessing Swedish foundries energy management program. *Energies* **2018**, *11*, 2780. [CrossRef]

54. DiMaggio, P.J.; Powell, W.W. The Iron Cage Revisited: Institutional Isomorphism and Collective Rationality in Organizational Fields. *Am. Sociol. Rev.* **1983**, *48*, 147–160. [CrossRef]

55. Helin, S.; Babri, M. Travelling with a Code of Ethics: A Contextual Study of a Swedish MNC Auditing a Chinese Supplier. *J. Clean. Prod.* **2015**, *107*, 41–53. [CrossRef]

56. Morris, T.; Lancaster, Z. Translating Management Ideas. *Organ. Stud.* **2006**, *27*, 207–233. [CrossRef]

57. Wæraas, A.; Sataøen, H.L. Trapped in conformity? Translating reputation management into practice. *Scand. J. Manag.* **2014**, *30*, 242–253. [CrossRef]

58. Cassell, C.; Lee, B. Understanding Translation Work: The Evolving Interpretation of a Trade Union Idea. *Organ. Stud.* **2017**, *38*, 1085–1106. [CrossRef]

59. Radaelli, G.; Sitton-Kent, L. Middle Managers and the Translation of New Ideas in Organizations: A Review of Micro-Practices and Contingencies. *Int. J. Manag. Rev.* **2016**, *18*, 311–332. [CrossRef]

60. Helin, S.; Sandström, J. Resisting a Corporate Code of Ethics and the Reinforcement of Management Control. *Organ. Stud.* **2010**, *31*, 583–604. [CrossRef]

61. Rolfsen, M.; Skaufel Kilskar, S.; Valle, N. "We are at day one of a new life": Translation of a Management Concept from Headquarter to a Production Team. *Team Perform. Manag.* **2014**, *20*, 343–356. [CrossRef]

62. Lillrank, P. The Transfer of Management Innovations from Japan. *Organ. Stud.* **1995**, *16*, 971–989. [CrossRef]

63. Doorewaard, H.; Van Bijsterveld, M. The Osmosis of Ideas: An Analysis of the Integrated Approach to IT Management from a Translation Theory Perspective. *Organization* **2001**, *8*, 55–76. [CrossRef]

64. Ragin, C. 'Casing' and the process of social inquiry'. In *What Is A Case? Exploring the Foundations of Social Inquiry*; Ragin, C., Becker, H., Eds.; Cambridge University press: Cambridge, UK, 1992.

65. Langley, A. Strategies for Theorizing from Process Data. *Acad. Manag. Rev.* **1999**, *24*, 691–710. [CrossRef]

66. Yin, R.K. *Case Study Research: Design and Methods*, 4th ed.; Sage: Los Angeles, CA, USA, 2009.

67. Mousavi, S.; Bossink, B.A.G. Firms' capabilities for sustainable innovation: The case of biofuel for aviation. *J. Clean. Prod.* **2017**, *167*, 1263–1275. [CrossRef]

68. Schofield, J.W. Increasing the generalizability of qualitative research. In *The Qualitative Researcher's Companion*; Huberman, M., Miles, M.B., Eds.; Sage Publication: Thousand Oaks, CA, USA, 2002.

69. Patton, M.Q. Two Decades of Developments in Qualitative Inquiry: A Personal, Experiential Perspective. *Qual. Soc. Work* **2002**, *1*, 261–283. [CrossRef]

70. Siggelkow, N. Persuasion with Case Studies. *Acad. Manag. J.* **2007**, *50*, 20–24. [CrossRef]

71. Eisenhardt, K.M.; Graebner, M.E. Theory Building from Cases: Opportunities and Challenges. *Acad. Manag. J.* **2007**, *50*, 25–32. [CrossRef]

72. Hsieh, H.F.; Shannon, S.E. Three approaches to qualitative content analysis. *Qual. Health Res.* **2005**, *15*, 1277–1288. [CrossRef]

73. Poole, M.S.; Ven, A.H.V.d. Empirical methods for research on organizational decision-making processes. In *Handbook of Decision Making*; Nutt, P.C., Wilson, D.C., Eds.; John Wiley: Chichester, West Sussex, UK; Hoboken, NJ, USA, 2010.

74. Røvik, K.A. From Fashion to Virus: An Alternative Theory of Organizations' Handling of Management Ideas. *Organ. Stud.* **2011**, *32*, 631–653. [CrossRef]

75. Harris, J.; Anderson, J.; Shafron, W. Investment in energy efficiency: A survey of Australian firms. *Energy Policy* **2000**, *28*, 867–876. [CrossRef]

76. Arens, M.; Worrell, E.; Eichhammer, W. Drivers and Barriers to the Diffusion of Energy-Efficient Technologies—A Plant-Level Analysis of the German Steel Industry. *Energy Effic.* **2017**, 1–17. [CrossRef]

77. Blass, V.; Corbett, C.J.; Delmas, M.A.; Muthulingam, S. Top Management and the Adoption of Energy Efficiency Practices: Evidence from Small and Medium-Sized Manufacturing Firms in the US. *Energy* **2014**, *65*, 560–571. [CrossRef]

78. Sandberg, P.; Söderström, M. Industrial Energy Efficiency: The Need for Investment Decision Support from a Manager Perspective. *Energy Policy* **2003**, *31*, 1623–1634. [CrossRef]

79. Solnørdal, M.T.; Thyholdt, S.B. Absorptive capacity and energy efficiency in manufacturing firms—An empirical analysis in norway. *Energy Policy* **2019**, *132*, 978–990. [CrossRef]

80. Johansson, M.T. Improved Energy Efficiency Within the Swedish Steel Industry—The Importance of Energy Management and Networking. *Energy Effic.* **2015**, *8*, 713–744. [CrossRef]

81. Burt, R.S. The Network Structure of Social Capital. *Res. Organ. Behav.* **2000**, *22*, 345–423. [CrossRef]

82. Balogun, J.; Hailey, V.H.; Gustafsson, S. *Exploring Strategic Change*; Pearson Education: London, UK, 2016.

83. Llewellyn, S. Two-way windows': Clinicians as medical managers. *Organ. Stud.* **2001**, *22*, 593–623. [CrossRef]

84. Raelin, J.D.; Cataldo, C.G. Whither middle management? Empowering interface and the failure of organizational change. *J. Chang. Manag.* **2011**, *11*, 481–507. [CrossRef]

85. Røvik, K.A. The secrets of the winners: Management ideas that flow. In *The Expansion of Management Knowledge: Carriers, Flows, and Sources*; Sahlin-Andersson, K., Engwall, L., Eds.; Stanford University Press: Palo Alto, CA, USA, 2002; pp. 113–144.

86. Apeaning, R.W.; Thollander, P. Barriers to and Driving Forces for Industrial Energy Efficiency Improvements in African Industries—A Case Study of Ghana's Largest Industrial Area. *J. Clean. Prod.* **2013**, *53*, 204–213. [CrossRef]

Publisher's Note: MDPI stays neutral with regard to jurisdictional claims in published maps and institutional affiliations.

MDPI

St. Alban-Anlage 66

4052 Basel

Switzerland

Tel. +41 61 683 77 34

Fax +41 61 302 89 18

www.mdpi.com

Sustainability Editorial Office

E-mail: sustainability@mdpi.com

www.mdpi.com/journal/sustainability

www.ingramcontent.com/pod-product-compliance
Lightning Source LLC
LaVergne TN
LVHW070627170726
843515LV00004B/194